3rd International Conference on Chemical Materials and Process (ICCMP 2018)

MATEC Web of Conferences Volume 187 (2018)

Bangkok, Thailand
23 – 25 May 2018

Editors:

J. Wang
Y.F. Yang

ISBN: 978-1-5108-6861-8

TABLE OF CONTENTS

PREFACE

2018 4th International Conference on Chemical Materials and Process (ICCMP 2018) was successfully held in Chulalongkorn University, Bangkok, Thailand during May 23-25, 2018. The objective of the Bangkok conference is to provide a platform for researchers, engineers, academicians as well as industrial professionals from all over the world to present their research results and development activities in Chemical Materials and Process.

ICCMP 2018 provided a chance for researchers, engineers and academicians to further develop Chemical Materials and Process. Attendees benefits a lot from expert practitioners and researchers from all over the world of the latest tendency in theoretical and practical domain of Chemical Materials and Process field. Distinguished professors also did their keynote speeches relating to the latest developments and trends in their respective fields. The keynote speakers included Prof. Sarawut Rimdusit from Chulalongkorn University, Bangkok, Thailand; Prof. Jian Chen from Tsinghua University, Beijing, China; Prof. Frank Yang from East Carolina University, Greenville, North Carolina, USA.

This set of conference proceedings presents a selection of papers submitted to the conference from a variety of universities, research institutes and industries. All papers were subjected to peer-review by conference committee members and a panel of international reviewers. The papers were selected to publish depend on the quality and relevance to the conference. This volume presents recent advances in the field of Multi-scale and/or Multi-disciplinary Approaches, Macromolecular Science and Engineering, Advanced materials processing, Process system, instrumentation and control, Chemical, Environmental, and Process Engineering, New materials & structured products, Chemistry and Chemical Engineering Fundamentals and etc.

We sincerely appreciate all the authors who have contributed to this volume and also to the organizing committee, reviewers, speakers, co-chair persons, sponsors and all the conference participants for their strong support to ICCMP 2018 to make this conference with a great succes. We look forward to meeting you in ICCMP 2019!

General Co-chair

Prof. Sarawut Rimdusit
Chulalongkorn University, Bangkok, Thailand
June 21, 2018

edp sciences

MATEC Web of Conferences

Statement of Peer review

In submitting conference proceedings to MATEC Web of Conferences, I certify to the Publisher that I adhere to the **Policy on Publishing Integrity** of the journal in order to safeguard good scientific practice in publishing.

1. All articles have been subjected to peer review administered by the proceedings editors.
2. Reviews have been conducted by expert referees, who have been requested to provide unbiased and constructive comments aimed, whenever possible, at improving the work.
3. Proceedings editors have taken all reasonable steps to ensure the quality of the materials they publish and their decision to accept or reject a paper for publication has been based only on the merits of the work and the relevance to the journal.

Title, date and place of the conference

2018 4th International Conference on Chemical Materials and Process (ICCMP 2018)
May 23-25, 2018
Chulalongkorn University, Bangkok, Thailand

Proceedings editor(s):

Prof. Jing Wang, University of South Florida, USA

Prof. Yu Frank Yang, East Carolina University, USA

Date and editor's signature 7/19/2018

Jing Wang 7/22/2018

Chemical Profiling Analysis and Identification the Bioactivities of Herbal Compress Extracts

Sumitta Chotikamas[1], Kraipat Cheenkachorn[1], Boochita Wongpanit[2], Prapakorn Tantayotai[3], and Malinee Sriariyanun[2]

[1]*Department of Chemical Engineering, King Mongkut's University of Technology North Bangkok, Bangkok 10800, Thailand*
[2]*The Sirindhorn International Thai-German Graduate School of Engineering, King Mongkut's University of Technology North Bangkok, Bangkok 10800, Thailand*
[3]*Department of Microbiology, Faculty of Science, Srinakarinwirot University, Bangkok 10110, Thailand*

Abstract. The traditional Thai herbal compress is composed of six different types of herbs including Ginger (*Zingiber cassumunar*), Turmeric (*Curcuma longa Linn.*), Soap Pod (*Acacia concinna*), Kaffir lime (*Citrus hystrix*), Lemongrass (*Cymbopogon citratus*), and Tamarind (*Tamarindus indica Linn.*). Herbal compress is used in treatment of Thai traditional massage to relieve the body pain and muscle strains. The objective of this work is to perform chemical profiling and extraction modelling of herbal compress obtained from solvent extraction method. The kinetic models, The second order and Fick's second law, representing the extraction behaviors of bioactive compounds were constructed to fit with experimental data of solvent extraction. Under tested condition, the extraction equilibrium was reached after 360 min and the second-order model gives the best fit for the experimental data with high coefficients of correlation (R^2 = 0.9927). Additionally, chemical profiling analysis showed that the amounts and variations of bioactive components in drying-pretreated herbal compress were more abundant than that of untreated sample. This finding could be applied further for preparation and production of traditional Thai herbal compress in the industrial scale.

1 Introduction

Herbal compress is made of combinations of various types of herbs that are tightly wrapped in a square piece of cloth. Before use, it is steamed in water for 15-20 min to allow the release of active compounds and aromatic oils from herb [1], and then placed directly on the skin to relieve pain and inflammation. Heat in steamed compress stimulates blood circulation and helps to treat sprains and pain. The benefits of using herbal compress in Thai traditional massage are to relax muscle and tendon, to increase body flexibility, to improve range of joint motion, and to enhance circulation of blood [2].

Generally, the main ingredients of Thai herbal compress include Ginger (*Zingiber cassumunar*), Turmeric (*Curcuma longa Linn.*), Soap Pod (*Acacia concinna*), Kaffir lime (*Citrus hystrix*), Lemongrass (*Cymbopogon citratus*), Tamarind (*Tamarindus indica Linn.*). Many researches have been investigated chemical compositions and biological properties of herbs. The main active chemicals of Ginger were identified to be sabinene, γ-terpinene, terpinen-4-ol and (E)-1-(3,4-dimethoxyphenyl) butadiene (DMPBD). DMPBD possesses a potent anti-inflammatory activity [3, 4]. Turmeric contains essential oils that possess various pharmacological activities, for example antioxidant, antimicrobial, anticancer and anti-inflammatory treatments, and the major chemicals, for example ar-turmerone, β-turmerone, arcurcumene, curlone, cubenol, β-sesquiphellandrene, germacrone and α-zingiberene [5-7]. Soap Pod helps in getting rid of skin infections and diseases due to its antibacterial properties [8]. Soap pod extracts has been demonstrated to have *in vitro* antioxidant and anti-tyrosinase activities [9]. Kaffir lime possesses antioxidant, antimicrobial, antibacterial properties and the major compounds are sabinene, β-pinene, limonene, α-pinene, camphene, myrcene, terpinen-4-ol, α-terpineol, linalool, terpinolene and citronellal [10-12]. The main compounds in the essential oil of lemongrass are the monoterpene citral, a mixtures of the stereisomers geranial , neral and β-myrcene and antioxidant properties has also been reported [13, 14]. Tamarind extract owns antioxidant and anti-inflammatory properties [15, 16].

Traditionally, Thai herbal compress could be made of either fresh herb or dried herb. The bioactivities and chemical compositions of the herbs are possibly loss during inappropriate pretreatment process, storage and usage of herbal compress. The database of bioactivities and chemical compositions of Thai herbal compress extracts is currently not yet completely collected. Therefore, the objective of this work is to perform chemical profiling and extraction modelling of herbal compress obtained from solvent extraction method. The extraction models were constructed based on experimental data that could be used to explain the phenomena to improve the production and Thai herbal compress in industrial scale.

2 Materials and methods

2.1 Materials

Ginger (*Zingiber cassumunar*), Turmeric (*Curcuma longa Linn.*), Soap Pod (*Acacia concinna*), Kaffir lime (*Citrus hystrix*), Lemongrass (*Cymbopogon citratus*), Tamarind (*Tamarindus indica Linn.*) were purchased from local market in Bangkok, Thailand. The plant materials were washed with distilled water, chopped into small pieces and dried in hot air oven at 60°C for 24 h. Solvents including 95%ethanol, hexane and water were selected for solid-liquid extraction of plant materials.

2.2 Solid-liquid extraction

Twenty grams of plant materials were extracted with 160 ml of each solvent for 0.5, 1, 2, 3, 4, 5, 6, 24 and 48 h at 30 °C. The extracts were filtered using filter paper (Whatman No.1).

2.3 UV-VIS Spectroscopic analysis

The UV-Visible absorption spectra of the samples were measured on a T80+ PG Instruments Ultraviolet–visible Spectrophotometer (UV-Vis) in the range of 400-700 nm of wavelength.

2.4 Gas chromatography–mass spectroscopy analysis

The extracted samples were analyzed by using GC–MS (GCMS-QP2020 Gas Chromatograph Mass Spectrometer, Shimadzu). GC–MS system equipped with a DB-WAX fused capillary column (30 m x 0.25 mm and 0.25 µm film thickness). Oven temperature was set at 50–250°C with ramping rate of 5°C/min, carrier gas was helium with a flow rate of 0.69 ml/min and spilt ratio 1:20.

2.5 Kinetic models

Two kinetic models were selected to fit experimental data, Second-order Equation and Fick's law Equation:

2.5.1 Second-order model

The second-order kinetic equation [17, 18] for the extraction rate can be written as follows:

$$\frac{dc_t}{dt} = k(c_\infty - c_t)^2 \quad (1)$$

Where k is the second-order extraction rate constant ($Lmol^{-1}min^{-1}$), C_∞ is the concentration of extracted at equilibrium ($molL^{-1}$), C_t is the concentration of extracted at any time ($molL^{-1}$) and t (min). Equation (2) is obtained by integrating with the initial and boundary conditions $C_t = 0$ at $t = 0$ and $C_t = C_t$ at $t = t$ and rearranging the following relations:

$$c_t = \frac{c_\infty{}^2 kt}{1 + c_\infty{}^2 kt} \quad (2)$$

Equation (2) may be rearranged to obtain the linear form:

$$\frac{t}{c_t} = \frac{1}{kc_\infty{}^2} + \frac{t}{c_\infty} \quad (3)$$

The initial extraction rate, as C_t /t when t approaches 0, can be defined as:

$$h = kc_\infty{}^2 \quad (4)$$

Equation (3) can be rearranged to finally obtain:

$$\frac{t}{c_t} = \frac{1}{h} + \frac{t}{c_\infty} \quad (5)$$

The initial extraction rate h, the concentration of extracted at equilibrium, C_∞ and the second-order extraction rate constant k can be determined experimentally from the slope and the intercept by plotting t/C_t versus t. Thus, the concentration of extracted at any time can be described as:

$$c_t = \frac{t}{(1/h) + (t/c_\infty)} \quad (6)$$

2.5.2 Fick's law model

From Fick's second law [19, 20] derived the applied diffusion model as follows:

$$\frac{\partial C}{\partial t} = \nabla(-D_{eff} \nabla C) \quad (7)$$

Where C is the concentration of solute, t is the time (s) and D_{eff} is the diffusivity or effective diffusion coefficient of the solute (m^2s^{-1}). The initial and boundary conditions for solving the equation were as follows:

$$t = 0,\ C = C_0 \quad \forall r$$

$$t > 0,\ C = C_i = 0 \quad r = R$$

$$t > 0,\ \frac{\partial C}{\partial r} = 0 \quad r = 0$$

For these the conditions and the assumptions of the mass transfer, Crank [19] proposed the solutions for diffusion in the radial direction (Eq. (8)).

$$\frac{C - C_0}{C_i - C_0} = 1 + [\frac{2R}{\pi r} \sum_{n=1}^{\infty} \frac{(-1)^n}{n} \sin \frac{\pi n r}{R} \exp\{-\frac{D_{eff} n^2 \pi^2 t}{R^2}\}] \quad (8)$$

The total amount of diffusing substance leaving the sphere as:

$$\frac{M}{M_\infty} = 1 - \frac{6}{\pi^2}\sum_{n=1}^{\infty}\frac{1}{n^2}\exp\{-\frac{D_{eff}n^2\pi^2 t}{R^2}\} \quad (9)$$

Where M_∞ is the total amount of solute transferred after infinite time, R is the radius of the sphere (m) and n is an integer number. The external mass transfer resistance is negligible [20, 21]. Equation (9) can be reduced to the following form:

$$\frac{M}{M_\infty} = \ln(\frac{6}{\pi^2}) - \frac{D_{eff}\pi^2 t}{R^2} \quad (10)$$

The concentration of the solute in the extraction solvent at any time following:

$$\ln\frac{C_\infty}{C_\infty - C} = 0.498 + \frac{9.87 D_{eff} t}{R^2} \quad (11)$$

Where C is the concentration of the solute in the extraction solvent at time *t*, C_∞ is the concentration of the solute in the extraction solvent after infinite time.

3 Results and discussion

3.1 Kinetic models

In this study, plants materials were extracted using the solid-liquid extraction method with three types of solvents as our previous study [22, 23]. Within 360-min-period observation, the concentration of the extract released from herbal compress was increased rapidly in early period, then later the concentration was gradually increase with slower rate suggesting the equilibrium of extraction. Similar observations were reported by Y.S. Ho. et al. [17] and H.S. Kusuma et al. [24]. This behavior could be explained that during the initial stage of extraction, when the solvent infiltrated into the solid, the high concentration gradient was presented, resulting in the high rates of mass transfer into the liquid phase. As the extraction time gaining, the mass transfer of the solute from solid phase to solvent was limited due to the reducing of concentration gradient.

The simplified mechanism of this behavior involves only two stages, namely washing stage (fast extraction) and diffusion stage (slow extraction) of extracted substances. As a rule, diffusion of extractable substance through plant particles is the rate limiting step of the overall extraction process because external resistance mass transfer is usually negligible [19, 20]. Therefore, in this study, two kinetic extraction models, Second-order model and Fick's law model were selected to fit with experimental data.

3.2 Second-order model

The comparisons between experimental extraction data and second-order rate models of all plant materials were shown in Figure 1. The concentration at equilibrium (C_∞), the initial extraction rate (h) and the second-order extraction rate constant (k) were calculated by plotting t/Ct versus t (time) as shown in Table 1.

The result of model fitting to experimental data showed that the Second-order model has average R^2 of 0.9927. Previously, R.R. Lovasoa *et al.* also applied the Second-order model to the Solid-liquid extraction system of *Fumaria officinalis* and obtained the highest R^2 value [25, 26].

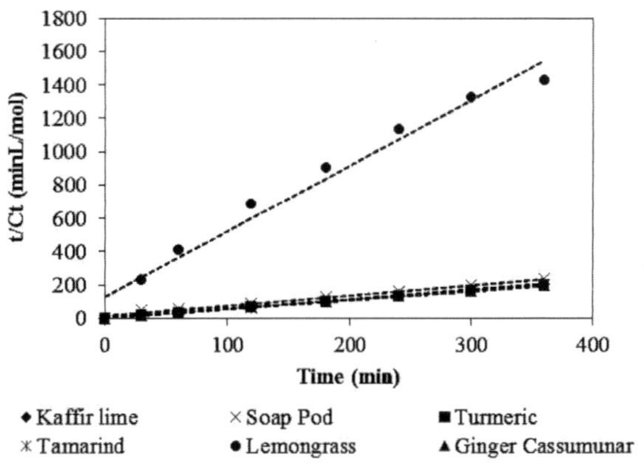

Figure 1. Second-order model represented extraction kinetics of extracts obtained from various plant materials.

Table 1. Linierization of second-order kinetic model of herbal compress extracts.

Plants	h (mol/L· min)	c_∞ (mol/L)	k (L/mol· min)	R^2
Kaffir lime	1.740	0.169	0.512	0.9998
Ginger Cassumunar	0.254	0.119	0.007	0.9750
Soap pod	1.818	0.576	1.908	1.0000
Turmeric	1.640	0.022	0.061	0.9820
Tamarind	1.832	0.247	0.832	0.9999
Lemongrass	1.827	0.356	1.191	0.9999
			Average	0.9927

3.3 Fick's law model

Next, Fick's law model was applied to find the correlation with the experimental data. This model is valid in the case of neglecting external mass transfer resistance. The predicted values were fitted to the experimental data and shown as a series of plots between $\ln(C_\infty/C_\infty\text{-}C)$ versus t (time) (Figure 2). The calculation of the diffusivity (D_{eff}) value obtained from the model was shown in Table 2.

It could be found that Fick's law model overlaid well, with average R^2 value of 0.9099, to the experimental data during the early phase of extraction time, especially during 360 min. Compared with the second-order extraction model, the second-order model presented the better correlation to experimental data. This finding could be further studied in more details to draw the proposed behaviors of extraction of herbal compress materials.

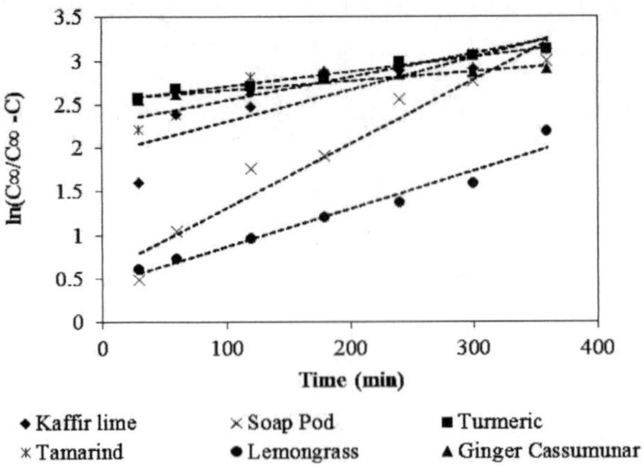

Figure 2. Fick's law model represented extraction kinetics of extracts obtained from various plant materials.

Table 2. Linearization of fick's law kinetic model of herbal compress extracts.

Plants	Diffusivity x 10^{12} (m^2/s)	R^2
Kaffir lime	3.7	0.7641
Ginger Cassumunar	4.3	0.9599
Soap pod	1.1	0.9343
Turmeric	7.3	0.9449
Tamarind	2.7	0.8819
Lemongrass	1.7	0.9748
	Average	0.9099

3.4 GC-MS analysis

The chemical profiles of herbal compress extracts obtained by using three types of solvents were acquired and analyzed with GC-MS as described in our previous studies [22, 23]. The identities of active compounds with their retention time (RT) and %abundance are shown in Table 3. Comparing chemical profiling obtained from 95%ethanol, hexane and water, the result showed that there are many chemicals identified in 95%ethanol and water extracts, but not in hexane extracts. Also, the variations of chemicals in hexane extracts are relatively less compared to 95%ethanol and water extracts.

In this study, many bioactive chemicals were identified differentially in each types of herbal materials. The main components of extracts from fresh and dried kaffir lime were elemol, δ-cadinene, 3-methyl-3-butenoic acid, acetic acid and α-terpineol. α-terpineol, which possesses antibacterial, anti-inflammatory properties antifungal activities [27-29], was present with high abundance in ethanol extracts at 10.55% and 6.92% of fresh and dry kaffir lime, respectively. The main components of extracts from fresh and dried lemongrass were 2-furanmethanol, acetic acid, 1-hydroxy-2-propanone, selina-6-en-4-ol (with 56.3% abundance and only presents in ethanol extract), and [R,R]-2,3-butanediol.

The main components of extracts from fresh and dried ginger were triquinacene,1,4-bis(methoxy) (with 70.27% abundance), terpinen-4-ol, and -Sesquiphellanderene. The main components of fresh and dried Soap pod extracts were furfural (only presented in water extract with 7.9% abundance), linoleic acid, 5-hydroxymethylfurfural. This result was confirmed by P. Sombatsiri et al. the compounds of *Acacia concinna (Wild.) DC* [30]. The main components of extracts from fresh and dried tamarind leaves were sabinene, β-pinene, α-pinene, D-limonene. The main components of extracts from fresh and dried turmeric were turmerone, β-sesquiphellanderene, terpinen-4-ol, β–bisabolene. It has been reported that the major compositions found in turmeric oil are α-phellanderen, terpinolen, α-zingiberene, r-turmerone, β-sequiphellanderen and α-turmerone [31].

The results indicated that ethanol and water were the most suitable solvents for extraction of herb from herbal compress. Interestingly, the traditional method to prepare herbal compress before use with consumer is done by using water steam. This study brought about the concern how to improve the effectiveness of compress for medicinal purpose. That is, to maximize the release of bioactive compounds from herbal compress, the mixed ethanol and water could be suggested.

Addtionally, the pretreatment method by drying of herbal materials has an effect on the profiles of remaining bioactive components, in both variation and abundance. Extracts with pretreatment with drying oven slightly gained abundance of major components compared to without pretreatment. This observation is similar to previous study that the essential oil of dried *C. citratus* leaves gave the highest amount of essential oil on dry weight basis [32].

4 Conclusion

In this study, the extracts of herbal compress were prepared by using solid-liquid extraction method and analyzed for chemical compositions. The influence of solvent type and pretreatment method were studied and the result showed that the pretreatment and solvent type have effects on the profiles of bioactive components of the herbal compress extracts. The kinetics of the solvent extraction were studied and correlated with two different extraction models. The results demonstrated that the second-order model achieved the best fit for the experimental data.

Acknowledgment

The authors would like to thank National Science and Technology Development Agency (NSTDA with STEM program), Srinakarinwirot University (Research grant No. 540/2560) and TF Cosmetology Co. LTD. for the financial support of this research.

Table 3. The major chemical components of herbal compress extracts.

	No	RT	Compound	% Abundance					
				95% Ethanol		Hexane		Water	
				Fresh	Dry	Fresh	Dry	Fresh	Dry
Kaffir lime	1	12.331	1-Hydroxy-2-propanone	-	-	-	-	6.03	8.65
	2	17.735	Acetic acid	3.92	-	-	-	24.21	26.07
	3	18.043	2-Propanone, 1-(acetyloxy)	-	-	-	-	2.29	3.28
	4	22.064	α-Terpineol	10.55	6.92	-	-	-	-
	5	23.011	δ-Cadinene	-	21.26	2.36	-	-	-
	6	23.752	3-Methyl-3-butenoic acid	13.46	-	-	-	20.11	22.72
	7	23.520	3-Methyl-2-butenoic acid	6.82	-	-	-	-	5.61
	8	28.063	Elemol	-	13.6	48.62	30.88	-	-
	9	29.255	(1R,7S,E)-7-Isopropyl-4,10-dimethylenecyclodec-5-enol	-	4.92	4.16	-	-	-
	10	29.854	α-Cadinol	-	-	5.19	-	-	-
Lemongrass	1	3.050	Formic acid	92.07	-	-	-	-	-
	2	12.368	1-Hydroxy-2-propanone	-	-	-	-	7.56	13.16
	3	16.116	Sulfurous acid, 2-ethylhexyl hexyl ester	-	-	4.95	13.8	-	-
	4	17.716	Acetic acid	-	-	-	-	27.93	20.25
	5	21.190	Hexacosane,1-iodo-	-	-	5.98	6.13	-	-
	6	21.447	2-Furanmethanol	0.79	-	-	-	3.84	1.31
	7	25.107	[R,R]-2,3-butanediol	-	-	-		-	53.96
	8	28.120	dl-Glyceraldehyde dimer	1.96	-	-	-	18.52	-
	9	29.372	Benzene, (1-propyloctyl)-	-	-	-	16.01	-	-
	10	29.557	Selina-6-en-4-ol	-	56.3	-	-	-	-
Ginger Cassumunar	1	3.021	Formic acid	22.88	-	-	-	-	-
	2	12.396	1-Hydroxy-2-propanone	-	-	-	-	10.15	-
	3	17.720	Acetic acid	-	-	2.27	-	27.86	30.13
	4	20.267	[R,R]-2,3-butanediol	-	-	-	-	-	62.58
	5	20.644	Terpinen-4-ol	27.33	3.1	-	-	-	-
	6	23.185	β-Sesquiphellanderene	9.53	2.61	21.55	-	-	-
	7	26.872	Triquinacene, 1,4-bis(methoxy)	-	70.27	-		-	-
	8	28.122	dl-Glyceraldehyde dimer	2.24	-	-	-	18.43	-
	9	29.503	2,2-dimethyl-4,5-di-1-propenyl-1,3-Dioxolane	2.87	-	-	-	-	0.35
	10	29.868	2-Methoxy-4-vinylphenol	12.3	2.28	-	-	-	0.92
Soap pod	1	11.941	Furfural	-	-	-	-	-	7.9
	2	16.382	2(5H)-Furanone	-	5.95	-	-	-	9.42
	3	21.369	1-Methyl-pyrrolidine-2-carboxylic acid	-	-	-	-	-	9.72
	4	22.818	5-Hydroxypipecolic acid	-	15.32	-	-	-	-
	5	24.228	Hexadecanoic acid, ethyl ester	-	16.57	-	-	-	-
	6	25.294	Propanoic acid	-	-	-	-	-	9.85
	7	27.557	Octadecanoic acid, ethyl ester	-	7.68	-	-	-	-
	8	28.103	5-Hydroxymethylfurfural	-	5.57	-	-	-	13.29
	9	28.647	Linoleic acid	-	5.41	-	-	-	-
	10	29.695	Ethyl linolenate	-	8.69	-	-	-	-
Tamarind	1	4.448	α-Pinene	-	-	-	3.54	-	-
	2	6.175	β-Pinene	-	7.23	-	35.41	-	-
	3	6.320	Sabinene	-	5.61	-	22.75	-	-
	4	8.538	D-Limonene	-	6.19	-	19.15	-	-
	5	17.890	3-Furaldehyde	-	-	-	-	-	26.49
	6	20.075	5-Methylfurfural	-	0.83	-	-	-	6.89
	7	23.107	1,2-Cyclopentanedione	-	-	-	-	-	4.21
	8	27.122	1H-Pyrrole-2-carboxaldehyde	-	-	-	-	-	4.32
	9	28.597	4,5-Dimethyl-1,3-dioxol-2-one	-	-	-	-	-	13.51
	10	29.437	5-Hydroxypipecolic acid	-	6.31	-	-	-	-
Turmeric	1	12.380	1-Hydroxy-2-propanone	-	-	-	-	11.91	14.84
	2	17.761	Acetic acid	-	-	-	-	53.89	57.13
	3	23.186	β-Sesquiphellanderene	7.76	-	0.78	-	-	-
	4	24.862	1,4(15),11-Eudesmatriene	1.06	9.18	-	1.45	-	
	5	25.500	Terpinen-4-ol	-	8.55	-	-	-	-
	6	27.311	7-epi-sesquithujene	-	16.31	-	26.23	-	-
	7	27.374	Beta.-Bisabolene			-	5.17	-	-
	8	28.043	β-Sesquiphellanderene	-		-	49.21	-	-
	9	29.403	Phenol, 2-methoxy-	-	34.02	-	-	-	-
	10	29.741	Turmerone	76.92	-	83.71	-	-	-

References

1. The Institute of Thai Traditional Medicine, The Public Manual of Health Care with Thai Traditional Medicine, The War Veterans Organization of Thailand Under Royal Patronage of His Majesty the King Publishing, Bangkok, Thailand, (2009)
2. W. Kumnerddee, Journal of the Medical Association of Thailand, **92**, 117 (2009)
3. U. Sukatta, P. Rugthaworn, P. Punjee, S. Chidchenchey, V. Keeratinijakal, Kasetsart Journal - Natural Science, **43**, 212 (2009)
4. R. Jeenapongsa, K. Yoovathaworn, K. M. Sriwatanakul, U. Pongprayoon, K. Sriwatanakul, J. Ethnopharmac, **87**, 143 (2003)
5. H. P. T. Ammon, M. A. Wahi, Planta Med, **57**, 1 (1991)
6. D.-W. Kim, S.-M. Lee, H.-S. Woo, J.-Y. Park, B.-S. Ko, J.-D. Heo, Y.-B. Ryu, W.-S. Lee, Journal of Functional Foods, **26**, 485 (2016)
7. L. Zhang, Z. Yang, F. Chen, P. Su, D. Chen, W. Pan, et al., Industrial Crops and Products, **109**, 60 (2017)
8. M. Wuthi-udomlert, O. Vallisuta, Pharmacogn J., **3**, 69 (2011)
9. W. Poomanee, W. Chaiyana, N. Intasai, P. leelapornpisid, Int J Pharm Pharm Sci, **7**, 237 (2015)
10. S. Siripongvutikorn, P. Thummaratwasik, Y. Huang, LWT-Food Science and Technology, **38**, 347 (2005)
11. V. Srisukh, C. Tribuddharat, V. Nukoolkarn, N. Bunyapraphatsara, K. Chokephaibulkit, S. Phoomniyom, et al., ScienceAsia, **38**, 212 (2012)
12. N. Kasuan, Z. Muhammad1, Z. Yusoff1, M. H. F. Rahiman, M.N. Taib, Z. A. Haiyee, MJAS, **17**, 359 (2013)
13. M R Santin, A O Santos ,C. Nakamura & B. P. D. Filho ,I. C. P. Ferreira, T. Ueda-Nakamura, Parasitol Res, **105**, 1489 (2009)
14. M. d. C. Vázquez-Briones, L. R. Hernández, J. A. Guerrero-Beltrán, J F R, **4**, 36 (2015)
15. SS. Bhadoriya, V. Mishra, S. Raut, A. Ganeshpurkar, SK. Jain, Sci Pharm, **80**, 685 (2012)
16. P. Kuru, APJTB, **4**, 676 (2014)
17. Y.-S. Ho, H.-O.H. Adamou, H. Fauduet, C. Porte, Separation and Purification Technology, **45**, 169 (2005)
18. H.-O.H. Adamou, H. Fauduet, C. Porte, Y.-S. Ho, Chemical Engineering Communications, **194**, 537 (2007)
19. J. Crank, *The Mathematics of Diffusion*, 2nd edition. Oxford, UK., (1975)
20. C.-H. Chan, R. Yusoff, G.-C. Ngoh, ChERD, **92**, 1169 (2014)
21. M. Pinelo, J. Sineiro, M. J. Nunez, Journal of Food Engineering, **77**, 57 (2006)
22. S. Chaichoowong, J.B. Bol, P. Bol, T. Gamse, M. Sriariyanun, Oriental Journal of Chemistry, **33**, 66 (2017)
23. C. Muenmuang, M. Narasingha, T. Phusantisampan, M. Sriariyanun, ACM International Conference Proceeding Series Part F130950, 119 (2017)
24. H.S. Kusuma, M. Mahfud, International Food Research Journal, **24**, 1697 (2017)
25. R-R. Lovasoa, J-L. Havet, C. Porte, H. Fauduet, Separation and Purification Technology, **54**, 253 (2007)
26. R-R. Lovasoa, J-L. Havet, C. Porte, H. Fauduet, Industrial Crops and Products, **29**, 516 (2009)
27. S. Held, P. Schieberle, V. Somoza, J.Agric. Food Chem, **55**, 8040 (2007)
28. R. Kotan, S. Kordali, A. Cakir, ZNC, **62**, 507 (2007)
29. D. Pitarokili, M. Couladis, P-P. Niki, O. Tzakou, J.Agric. Food Chem, **50**, 6688 (2002)
30. P. Sombatsiri, G. Chairote, *In Proc. WOCMAP III,: Quality, Efficacy, Safety, Processing & Trade in MAPs.*, 189 (2005)
31. G. Asghari, A. Mostajeran, M. Shebli, RPS, **4**, 55 (2009)
32. A.R. Mohamed Hanaa, Y.I. Sallam, A.S. El-Leithy, Safaa E. Aly, AOAS, **57**, 113 (2012)

MATEC Web of Conferences **187**, 01002 (2018) https://doi.org/10.1051/matecconf/201818701002
ICCMP 2018

Study of Mathematical Models in Hot Air Drying of Herbs in Herbal Compress Ball

Boochita Wongpanit[1], Sumitta Chotikamas[2], Supacharee Roddecha[3], Prapakorn Tantayotai[4] and Malinee Sriariyanun[1]

[1] *The Sirindhorn International Thai-German Graduate School of Engineering, King Mongkut's University of Technology North Bangkok, Bangkok 10800, Thailand*

[2] *Department of Chemical Engineering, Faculty of Engineering, King Mongkut's University of Technology North Bangkok, Bangkok 10800, Thailand*

[3] *Department of Chemical Engineering, Faculty of Engineering, Kasetsart University, Bangkok 10900, Thailand*

[4] *Microbiology Department of Microbiology, Faculty of Science, Srinakarinwirot University, Bangkok 10110, Thailand*

Abstract. Herbal compress ball is currently one of important products of Thailand for exporting sales worldwide. It is used in Thai traditional medical treatment and spa to reduce muscle pain and relaxation. This research aimed to generate the mathematical models representing the behaviors of herbs in hot air drying to extend shelf life for exporting sales. Here, six types of herbs, including Prai (*Zingiber cassumunar* Roxb.), Turmeric (*Curcuma longa* Linn.), Lemongrass (*Cymbopogon citratus*), Kaffir lime (*Citrus hystrix*), Soap Pod leaves (*Acacia concinna*) and Tamarind Leaves (*Tamarindus indica* Linn.) were dried in different temperature at 60, 70, and 80 °C. Fours drying models, Page, Henderson and Pabis, and Logarithmic and Fick's second law equation were applied with experimental data of drying herbs to predict the rate of diffusion of water. The results showed that the Page model is the most suitable model due to the highest decision coefficient (R^2) but the lowest Root Mean Square Error (RMSE). The effective moisture diffusivity (D_{eff}) of the herbs in herbal compress ball was increased with increased the drying temperature. The size of the herb particle translated inversely with effective moisture diffusivity (D_{eff}) value.

1 Introduction

Based on data from the World Health Organization (WHO), 80 % of the population in developing countries have used traditional medicine and herbs for treatment of sickness and nourishing health. Applications of herbs for these purposes has been inherited and developed for a long time [1]. One of applications of herbs in Thailand is in the form of herbal compress, so called Luk-Prakob in Thai, that various types of mixed herbs are wrapped in the thin cheese clothes. Thai herbal compress is mainly used in muscle pain treatment and in spa [2]. The medicinal effects of herbal compress depend on the composition of herbs in the compress.

One of the traditional formula of Thai compress consists of 6 types of herbs, including Prai (*Zingiber cassumunar* Roxb.), Turmeric (*Curcuma longa* Linn.), Lemongrass (*Cymbopogon citratus*), Kaffir lime (*Citrus hystrix*), Soap Pod leaves (*Acacia concinna*), and Tamarind Leaves (*Tamarindus indica* Linn.). Several previous studies reported that Prai (*Zingiber cassumunar* Roxb.), kaffir lime (*Citrus hystrix*) [3], Tamarind Leaves (*Tamarindus indica* Linn.) [4], and turmeric (*Curcuma longa* Linn.) [5] have anti-inflammatory activities. Lemongrass (*Cymbopogon citratus*) posesses anti-oxidant, anti-microbial and anti-fungal activities [6]. Soap Pod leaves (*Acacia concinna*) has major compositions, such as phenolics, flavonoids, and carotenoids that have antioxidant activities [7].

Herbal compress product has one important limitation of storage because herbs in the herbal compress ball have high moisture causing the growth of microorganisms and spoilage. Therefore, the drying process is a necessary process in preserving the product for a longer shelf life. By reducing moisture or water content in the product to inhibit the microorganisms that cause the spoilage products [8]. Hot air drying is the most common technique in the industry by providing hot air flow through wet material.

The principle of drying process involves heat and mass transfer [9]. Mathematical Modelling is a method for prediction the change of moisture content in the product during drying in the experiment. The information obtained from the model such as initial moisture content, drying time, temperature, and energy values could be used to design a better efficient dryer to save time and cost [10-12]. However, each drying model still has limitation that could be applied in different experimental scenario. In this research, we focused on studying the drying process of the herbs in herbal compress ball with a hot air dryer to select the appropriate mathematical model that fit well to the experimental data.

2 Materials and methods

2.1. Material preparation

The fresh plant materials, including Prai (*Zingiber cassumunar* Roxb.), Turmeric (*Curcuma longa* Linn.), Lemongrass (*Cymbopogon citratus*), Kaffir lime (*Citrus hystrix*), and Tamarind Leaves (*Tamarindus indica* Linn.), were purchased from the local market in Bangkok, Thailand. The samples were washed with distilled water, dried with ambient air (approximately 30 °C) and were prepared into 2 sizes, uncut and cut samples, then followed by size screening with sieve shakers at mesh no. 10 (avg. size 1.5 mm), and stored in a refrigerator at 4 ± 0.5 °C before the experiment.

2.2. Drying system and procedure

A hot-air convention dryer was used. The characteristic of dryer chamber is 370 mm × 340 mm × 420 mm. The air flow was forced through to chamber by using a blower in horizontal direction and then was heated as it passed through coil heaters that had a heat capacity of 900 Watt, operated at 230 V, 1 phase, and 60 HZ. The drying of samples were operated at 3 temperature levels are 60, 70, 80 ºC. After the dryer conditions were adjusted for the experiment plan then 200 g of plant materials was place in dryer and measure the weight of the remaining samples from drying every 30 minutes until constant weight.

2.3. Mathematic model and drying rate

The selected mathematical models used in this study are shown in Table 1. By the moisture ratio (MR) of herbs in Thai herbal compress ball during drying was calculated by using Eq. 1.

$$MR = \frac{Mt - Meq}{Mi - Meq} \tag{1}$$

where M_t is the moisture content at any time of herbs in Thai herbal compress ball, M_i is the initial moisture content of herbs in Thai herbal compress ball, and M_{eq} is the equilibrium moisture content of herbs in Thai herbal compress ball. For the dried agricultural products, the equilibrium moisture content can be assumed to be zero because the values of M_{eq} are relatively low when compared to M_t or M_i [13]. For this reason, the moisture ratio was calculated by using Eq. 2.

$$MR = \frac{Mt}{Mi} \tag{2}$$

The mathematical models (Table 1) used to find the optimal model for describing the drying curve of herbs in Thai herbal compress ball. Non-linear regression was used to achieve each constant parameter of selected mathematical models by using the MATLAB program for selecting the best model for describing the drying curves of the experiment.

2.4. Statistical analysis for determination of appropriate model.

Table 1. The mathematical models used in this study.

No.	Model name	Model equation	Reference
1	Page	$MR = \exp(-kt^n)$	[14]
2	Henderson and Pabis	$MR = a.\exp(-kt)$	[15]
3	Logarithmic	$MR = a.\exp(-kt) + c$	[16]

where a, k, c and n are constant parameter; t is drying time (sec).

The correlation coefficient (R^2) and root mean square error (RMSE) were used to evaluate the best mathematical model for drying herb in Thai herbal compress ball. The best value of correlation coefficient (R^2) is high (closer to one) and the value of root mean square error (RMSE) is low (closer to zero) [17]. These two parameters could be calculated by using Eq. (3)-(4):

$$R^2 = 1 - \frac{\sum_{i=1}^{N} \cdot (MR_{pre,i} - MR_{exp,i})^2}{\sum_{i=1}^{N} \cdot (\overline{MR}, \exp - MR_{pre,i})^2} \tag{3}$$

$$RMSE = \sqrt{\frac{1}{N}\sum_{i=1}^{N}(MR_{pre,i} - MR_{exp,i})^2} \tag{4}$$

where $MR_{exp,i}$ is the ith experimental value, $MR_{pre,i}$ is the ith predicted value, $MR_{exp,ave}$ is the ith experimental average value, N is the total number of observations.

2.5. Effective moisture diffusivity.

Fick's law of Diffusion was used to describe the water movement of herbs in the herbal compress ball during drying. The migration rate of water molecules or moisture within herb materials could be directly affected by the shapes of materials [18-20]. The equation can be expressed by Eq. 5- Eq. 7.

In this study, Prai (*Zingiber cassumunar* Roxb.), Turmeric (*Curcuma longa* Linn.), and Lemongrass (*Cymbopogon citratus*), were assumed to be cylindrical to find the effective moisture diffusivity (D_{eff}). And the equation for a cylindrical shape can be expressed by Eq.5

$$MR = \frac{Mt - Meq}{Mi - Meq} = 4.\exp(-\frac{Deff.t}{R^2}) \tag{5}$$

For Tamarind Leaves (*Tamarindus indica* Linn.) and Kaffir lime (*Citrus hystrix*) were assumed to be infinite slab to find the effective moisture diffusivity (D_{eff}). Equation can be expressed by Eq.6

$$MR = \frac{Mt - Meq}{Mi - Meq} = \frac{8}{\pi^2}\exp(-\pi^2\frac{Deff.t}{4L^2}) \tag{6}$$

and the equation for a spherical shape was described the reduced size material as show in Eq. 7

Table 2. Statistical analysis with the Page, Henderson and Pabis and logarithmic model on drying process.

Drying model	Sample	Temperature (°C)	Drying Model Constants	R^2	RMSE
Page	Prai (*Zingiber cassumunar* Roxb.)	60	k = 0.000304, n = 1.365	0.9974	0.01519
		70	k = 0.004204, n = 1.000	0.9989	0.01045
		80	k = 0.003291, n = 1.061	0.9910	0.02812
	Turmeric (*Curcuma longa* Linn.)	60	k = 0.000477, n = 1.262	0.9969	0.01584
		70	k = 0.004542, n = 1.012	0.9974	0.01519
		80	k = 0.001706, n = 1.221	0.9996	0.00627
	Lemongrass (*Cymbopogon citratus*)	60	k = 0.000306, n = 1.386	0.9987	0.01141
		70	k = 0.000635, n = 1.327	0.9962	0.01993
		80	k = 0.002676, n = 1.079	0.9966	0.01770
	Kaffir lime (*Citrus hystrix*)	60	k = 0.001765, n = 1.262	0.9984	0.01341
		70	k = 0.002703, n = 1.219	0.9993	0.00864
		80	k = 0.003286, n = 1.258	0.9991	0.00967
	Tamarin (*Tamarindus indica* Linn.)	60	k = 0.002149, n = 1.385	0.9981	0.01618
		70	k = 0.010070, n = 1.150	0.9970	0.01931
		80	k = 0.004381, n = 1.322	0.9913	0.03422
Henderson and Pabis	Prai (*Zingiber cassumunar* Roxb.)	60	a = 1.011, k = 0.004919	0.9975	0.01478
		70	a = 0.047, k = 0.006708	0.9965	0.01870
		80	a = 1.013, k = 0.008602	0.9908	0.02828
	Turmeric (*Curcuma longa* Linn.)	60	a = 1.047, k = 0.002333	0.9873	0.03191
		70	a = 1.011, k = 0.004919	0.9975	0.01478
		80	a = 1.061, k = 0.005880	0.9937	0.02591
	Lemongrass (*Cymbopogon citratus*)	60	a = 1.084, k = 0.003069	0.9841	0.03939
		70	a = 1.077, k = 0.004226	0.9842	0.04094
		80	a = 1.026, k = 0.004256	0.9959	0.01950
	Kaffir lime (*Citrus hystrix*)	60	a = 1.059, k = 0.007230	0.9890	0.03474
		70	a = 1.045, k = 0.008456	0.9924	0.02843
		80	a = 1.043, k = 0.011540	0.9911	0.03114
	Tamarin (*Tamarindus indica* Linn.)	60	a = 1.056, k = 0.013210	0.9815	0.05036
		70	a = 1.012, k = 0.019180	0.9944	0.02649
		80	a = 1.028, k = 0.017870	0.9811	0.05047
Logarithmic	Prai (*Zingiber cassumunar* Roxb.)	60	a = 0.998, k = 0.005178, c = 0.01987	0.9981	0.01328
		70	a = 1.052, k = 0.006540, c = -0.00945	0.9967	0.01874
		80	a = 1.002, k = 0.009182, c = 0.02055	0.9925	0.02645
	Turmeric (*Curcuma longa* Linn.)	60	a = 1.122, k = 0.002031, c = -0.08468	0.9908	0.02797
		70	a = 0.998, k = 0.005178, c = 0.01987	0.9981	0.01328
		80	a = 1.085, k = 0.005429, c = -0.03345	0.9955	0.02257
	Lemongrass (*Cymbopogon citratus*)	60	a = 1.138, k = 0.002752, c = -0.06361	0.9869	0.03694
		70	a = 1.112, k = 0.003869, c = -0.04453	0.9862	0.03944
		80	a = 1.031, k = 0.004188, c = -0.00721	0.9959	0.01995
	Kaffir lime (*Citrus hystrix*)	60	a = 1.102, k = 0.006166, c = -0.06335	0.9952	0.02365
		70	a = 1.079, k = 0.007314, c = -0.05324	0.9974	0.01719
		80	a = 1.067, k = 0.010420, c = -0.03623	0.9948	0.02483
	Tamarin (*Tamarindus indica* Linn.)	60	a = 1.120, k = 0.010730, c = -0.08503	0.9918	0.03558
		70	a = 1.012, k = 0.019190, c = 1.427E-13	0.9944	0.02649
		80	a = 1.063, k = 0.015870, c = -0.04438	0.9861	0.04631

Table 3. Effective moisture diffusivity at different temperature.

Temperature (°C)	Effective Moisture Diffusivity (D_{eff}) [m^2/s]									
	Prai		Turmeric		Lemongrass		Tamarin		Kaffir lime	
	Uncut size	Cut size (avg.1.5 mm)	Uncut size	Cut size (avg.1.5 mm)	Uncut size	Cut size (avg.1.5 mm)	Uncut size	Cut size (avg.1.5 mm)	Uncut size	Cut size (avg.1.5 mm)
60	1.10E-03	5.30E-04	6.82E-05	8.86E-05	7.11E-05	4.84E-06	2.86E-04	1.59E-04	4.37E-04	8.93E-04
70	1.29E-03	1.42E-02	8.26E-05	9.04E-05	7.85E-05	1.62E-04	4.37E-04	7.96E-02	5.34E-04	4.68E-01
80	1.32E-03	1.93E-02	8.97E-05	1.44E-04	8.02E-05	2.52E-04	4.25E-04	1.02E-01	7.76E-04	6.95E-01

$$MR = \frac{Mt - Meq}{Mi - Meq} = \frac{6}{\pi^2}.\exp(-\pi^2 \frac{Deff.t}{R^2}) \qquad (7)$$

where D_{eff} is effective moisture diffusivity in m^2/s, L is the thickness of sample in infinite slab equation (m), t is drying time (sec), and R is the radius of the sample in cylinder equation.

3 Results and discussion.

3.1. A drying characteristics of the herbs in herbal compress balls.

Based on the experimental data, the values of moisture ratio with a drying time of 5 different types of herbs at 60-80 °C were demonstrated in Figure 1a-1c. Drying behavior of herb materials showed the moisture in samples rapidly reduced in the early phase of drying, and the drying rate was reduced at the later phase until the moisture equilibrium was achieved [21]. In this study, among all types of herbs tested, Prai (*Zingiber cassumunar* Roxb.) has the lowest moisture reducing rate at 60 °C, following by Turmeric (*Curcuma longa* Linn.), Lemongrass (*Cymbopogon citratus*), Kaffir lime (*Citrus hystrix*), and Tamarind Leaves (*Tamarindus indica* Linn.), respectively (Figure 1a). At 70 °C, Lemongrass (*Cymbopogon citratus*) has the lowest moisture reduction rate, following by Prai (*Zingiber cassumunar* Roxb.), Turmeric (*Curcuma longa*), Kaffir lime (*Citrus hystrix*) and Tamarin (*Tamarindus indica* Linn.), respectively. However, after 300 minutes of drying time, Prai (*Zingiber cassumunar* Roxb.) turned to have the lowest moisture reduction rate instead. At 80 °C, Lemongrass (*Cymbopogon citratus*) has a lowest moisture reduction rate, following by Prai (*Zingiber cassumunar* Roxb.), Turmeric (*Curcuma longa* Linn.), Kaffir lime (*Citrus hystrix*), and Tamarind Leaves (*Tamarindus indica* Linn.), respectively.

3.2 Drying model

In this study, three different mathematical models including the Page, Henderson and Pabis, and Logarithmic were applied to predict the reduction of moisture content during drying of plant materials as shown in Table 2. The reliability of each model was evaluated based on correlation coefficient (R^2), and root mean square error (RMSE) value at 3 different temperatures at 60, 70, and 80 °C.

The results showed that Page model provided the best fit to the experimental drying data of Prai (*Zingiber cassumunar* Roxb.), turmeric (*Curcuma longa* Linn.), lemongrass (*Cymbopogon citratus*), kaffir lime (*Citrus hystrix*), and tamarind Leaves (*Tamarindus indica* Linn.) with R^2 values of 0.9957, 0.9979, 0.9971, 0.9989, and 0.9954, respectively and RMSE values of 0.01792, 0.0124, 0.0163, 0.0105, and 0.0232 respectively. The large value of R^2 (closer to 1.0) means the model fit well with the drying dataset. The small RMSE value means the smallest error of the generated model. Here, the Page model provided the highest R^2 and the lowest RMSE values suggesting that this model should be selected to represent the experimental data.

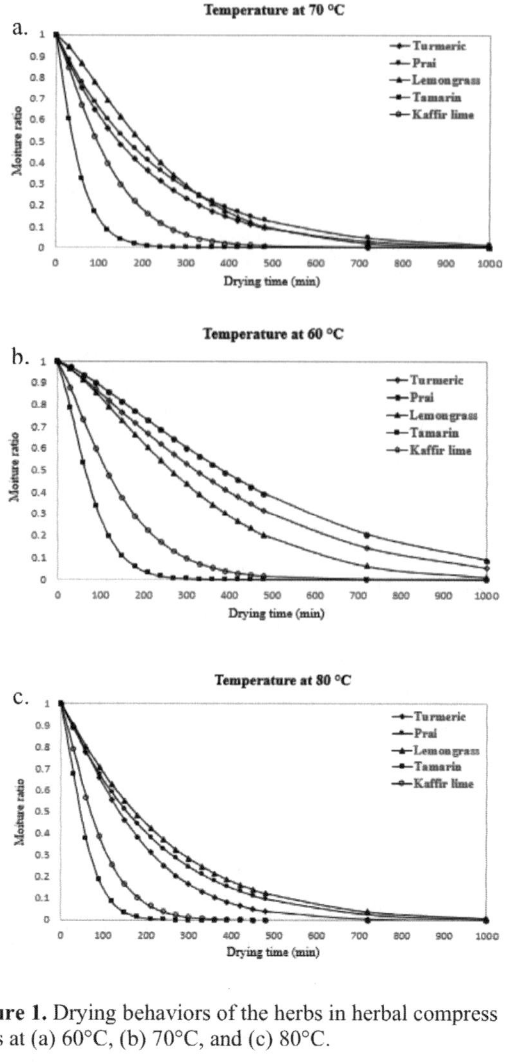

Figure 1. Drying behaviors of the herbs in herbal compress balls at (a) 60°C, (b) 70°C, and (c) 80°C.

3.3 Effective moisture diffusivity (D_{eff})

During the drying process of herbs material, the drying rate could be determined by observation of the decreased rate of moisture content within a certain time period. The moisture transfer in the drying process is controlled by internal diffusion. The effective moisture diffusivity (D_{eff}) could be explained by Fick's second law equation to describe the diffusion of water [22].

In this study, the values of effective moisture diffusivity (D_{eff}) at different temperatures and particle sizes of herbs were determined. The calculated D_{eff} values of each experiment were reported in Table 3. It could be observed that the higher drying temperature yielded the higher D_{eff} values. For the particle size or the size reduction effect, the D_{eff} values was higher in cut sample (smaller particle size). For example, at 80 °C, Prai (*Zingiber cassumunar* Roxb.) has effective moisture diffusivity (D_{eff}) value of 1.32×10^{-3} and 1.93×10^{-2} m^2/s for large size (uncut sample) and small size (cut sample). This observation suggested that the size reduction increased D_{eff} values because the small particle has the larger surface area than the big particle [23, 24].

4 Conclusion

The results showed that the drying temperature have an influence on drying behavior of herbs. The rate of reduction of moisture ratio was found to increase with increase the temperature suggesting that drying at higher temperature could reduce the herb preparation in compress ball. Here, Page model is the most fit to the experimental data compared to others model. In addition, the particle size of herb materials affected to the effective moisture diffusivity (D_{eff}) values, that is, the reduction of particle size increase the diffusion of water. Therefore, the knowledge of this study could be further applied to the production and preparation process for commercial herbal compress for export sale.

5 Acknowledgment

The authors would like to thank National Science and Technology Development Agency (NSTDA with STEM program), Srinakarinwirot University (Research grant No. 540/2560) and TF Cosmetology Co. LTD. for the financial support of this research.

References

1. P. Yasarin, M. Sriariyanun, and T. Phusantisampan, KMUTNB Int J Appl Sci Technol, **9** (2016)
2. N. Chiranthanut, N. Hanprasertpong, and S. Teekachunhatean, Hindawi Publishing Corporation BioMed Research International, vol. 2014, (2014)
3. W. A. S. Tunjunga, J. Cinatl jr. , M. Michaelisc, C. M. Smales, Procedia Chemistry, **14** (2014)
4. P. Kuru, Asian Pacific Journal of Tropical Biomedicine, **4** (2014)
5. Y. Zhang, X. Jiang, K. Peng, C. Chen, L. Fu, Z. Wang, J. Feng, Drug Des Devel Ther, **8** (2014)
6. M. Ranitha, Abdurahman H. Nour, Ziad A. Sulaiman, Azhari H. Nour, and Thana Raj S., International Journal of Chemical Engineering and Applications, **5** (2014)
7. W.Poomanee, W. Chaiyana, N.intasai, P. Leelapornpisid, Article in International Journal of Pharmacy and Pharmaceutical Sciences, **7** (2015)
8. Z. Erbay and F.Icier, Critical Reviews in Food Science and Nutrition, **50** (2009)
9. A.Tahmasebi, J. Yu, X. Li, C. Meesri, Fuel Process. Technol. ,**92** (2011)
10. A.T. Garavanda, S. Rafieea, A. Keyhania, Journal of Engineering, Management, & Applied Sciences & Technologies, **2** (2011)
11. T.O. Olurin, A. O. Adelekan, W. A. Olosunde, Agric Eng Int: CIGR Journal, **14** (2012)
12. E. N. Wami and M. O. Ibrahim, International Journal of Scientific & Engineering Research. **5** (2014)
13. M. Aghbashlo, MH. Kianmehr, S. Khani, M.Ghasemi, Int. Agrophysics, **11** (2017)
14. P. Gupta, J. Ahmet, U.S. Shivhare, G.S.V. Raghavon Drying technology, **20** (2007)
15. R. Kabganian, D.J. Carrier, S. Sokhansanj, Drying Technology, **20** (2007)
16. S. J. Temple, A. J. B. Boxtel, Journal of Agriculture and Engineering Research, **74** (1999)
17. Z. Erbay, F. Icier, Critcal Reviews in Food Science and Nutrition, **50** (2010)
18. N. Sinha,Y. H. Hui, E. Özgül Evranuz, M. Siddiq, J. Ahmed, Handbook of Vegetables and Vegetable Processing, (2010)
19. P. Simha, M. Mathew, M. Ganesapillai, Journal of Alexandria Engineering, **55** (2016)
20. B.D.A. Porciuncula, M.F. Zotarelli, B.A.M. Carciofi, J.B. Laurindo, Journal of Food Engineering, **119** (2013)
21. A.S. Mujumdar, and A.S. Menon, Drying of Solids: Principles, Classification, and Selection of Dryers. In: Handbook of Industrial Drying, 2nd ed. New York. Marcel Dekker Inc., (1995)
22. F.M. Botelho, P.C. Corrêa, A. L. D. Goneli , M. A. Martins, F.E.A. Magalhães and S. C. Campos, Revista Brasileira de Engenharia Agrícola e Ambiental, **15** (2011)
23. K.Y. Kone, Z.E. Gnimpieba, D. Soro, N.E. Assidjo, J.C. Laguerre, Asian Journal of Agriculture and Food Sciences, **5** (2017)
24. S. Chaichoowong, J.B. Bol, P. Bol, T. Gamse, M. Sriariyanun, Oriental Journal of Chemistry, **33**, 66 (2017)

 https://doi.org/10.1051/matecconf/201818701003

Investigating the Potential Use of Ionic Liquids in Pre-Treatment Application for Water Desalination

Hadil Abu Khalifeh[1], Inas AlNashef[2], Asiya Nazir[1], Noora Al Mansoori [1], Ibrahim Yousef [1] and Awatif Sharhahabiel [1]

[1] *Abu Dhabi University, Chemical Engineering Department, UAE*
[2] *Khalifa University, Department of Chemical and Environmental Engineering, UAE*

Abstract. The world has been noticing a quickening advancement over the last decades in terms of irrigated agriculture, social, industrial, and economical perspectives; followed by a huge increment in the water demand. Therefore, desalination is used all over the world to reduce worldwide water shortage; however, the traditional techniques lead to fossil fuel depletion and global warming. Therefore, scientists are investigating new green and environment-friendly methods to be used by the desalination plants to reduce CO_2 emissions and save the natural resources. In this study, the feasibility of using ionic liquids (ILs) as potential means for pre-treatment of seawater desalination was examined. The main aim of this work was to examine the ability of ILs in extracting salts from saline water. As a first step, the solubility of NaCl, $MgCl_2$, and $CaCl_2$ in different ILs at different temperatures were analyzed. The solubility of the salts in ILs increased in most cases with temperature increments; however, the presence of emulsion was seen in a few cases. The highest measured concentration of NaCl was 6,639 ppm at 60 °C in 1,3-Dimethylimidazolium dimethyl phosphate.

1 Introduction

Water occupies 79% of the earth's volume; nonetheless, only 2.5% of the total water is fresh water. The population is growing dramatically over the years and so is the demand on fresh water. Moreover, the expansion of irrigated agriculture and industrial advancement play a major role besides the population growth in increasing fresh water demand. Hence, saline water needs to be desalinated in order to provide more fresh water for the next generations. The typical method used in the distillation plants is the sea water reverse osmosis; however, to produce 1 m^3 of fresh water, 3-10 kW of electric energy is required [1]. The typical desalination plants depend on fossil fuels to afford this amount of energy which leads to increased fossil fuel depletion and CO_2 emission. Nowadays, finding a clean method for desalinating water is a top priority.

In this study, ILs will be evaluated to be used as means in the pre-treatment process in desalination plants. Even though ILs were initially defined in 1914, a noticeable consideration has been focused on utilizing ILs as a replacement to the traditional hazardous solvents in recent years [2, 3]. This was due to their unique and safe characteristics, such as, the feasibility of reusing, low vapor pressure, and low flammability [4]. This study is composed of two stages: Firstly, the solubility of single three different salts (NaCl, $MgCl_2$, and $CaCl_2$) in different ILs is tested at five different temperatures (25, 45, 60, 80 and 100°C). Secondly, the solubility of combined three different salts (NaCl, $MgCl_2$, and $CaCl_2$) in different ILs is measured at five different temperatures (25, 45, 60, 80 and 100°C). The results obtained in this stage are taken as a reference for extracting salts from saline water.

A recent study suggested that the presence of inorganic salts in ILs increases the ionicity of the ILs as salts empower the charge density of the IL [5]. A previous work was conducted for evaluating the solubility of NaCl in different ILs at different temperatures, and it was concluded that the solubility of NaCl and temperature have a proportional relationship and that NaCl is more soluble in imidazolium-based ILs than pyrrolidinium- and ammonium based ILs [6]. A recent study investigated the solubility of NaCl in different phosphonium-based DESs as a function of temperature and the salt: HBD (hydrogen bond donor) or metal halide molar ratio. The results show that increasing the temperature and the amount of HBDs or metal halides in the DES increases the solubility of NaCl in the DESs. The chemical structure of the DES significantly affects the solubility of NaCl in the studied DESs. The study tested the ability of different Trioctyl ammonium based ILs in extracting NaC1 from water, extraction was higher in a small anion based ILs because NaCI could easily diffuses between the cation and anions of the IL. Meanwhile with large anion based ILs the diffusion of NaCl is less; hence, lower extraction efficiency [7]. The benefit of using hydrophobic ILs in extraction is due to the elimination of water and IL cross-contamination. The hydrophobic ILs consist of the imide anion which contributes to their hydrophobic nature. The parameters to study are: temperature, type of salt, water to IL mass ratio, and IL structure. The results in this stage determine the feasibility of using ILs as part of the pre-treatment processes in desalination plants.

If the hypothesis is approved, then more fresh water will be produced from environmental friendly methods. This will subsequently diminish the salinity of seawater and decrease the stack on desalination plants; hence, the vital necessities will be reduced. Furthermore, this will significantly lessen the fouling and scaling of membranes generated by the salts' deposition, leading to decreased expenses as less maintenance services will be required and resources will be saved.

2 Methodology

2.1 Materials

Three different salts were used in this study namely, sodium chloride, magnesium chloride, and calcium chloride. The salts were supplied by Sigma-Aldrich. The specified salts were selected because of their significant concentration in seawater. Furthermore, ILs used in this study were purchased from Ionic Liquid Technologies, IoLiTec, (Heilbronn, Germany). Each ionic liquid from the selected ILs had different cation and/or anion. Also, some ILs were hydrophilic and the others were hydrophobic. The hydrophilic IL was only used in the solubility study alongside with one hydrophobic IL. Table 1 shows two of the ILs used in the solubility measurements part.

Table 1. Ionic Liquids used for salt solubility experiments.

IL Number	IL Name
14	1,3-Dimethylimidazolium dimethyl phosphate
15	Methyltrioctylammoniumbis (trifluoromethylsulfonyl) imide

2.2 Equipment

An analytical balance was used to measure the required mass of the salts. Eppendorf Thermomixer® C (Germany) was used to heat and mix the samples to the desired temperature. The concentration of the salts in ILs was determined by ICP-OES (Inductively Coupled Plasma-Optical Emission Spectrometry). Lastly, the water content of the ILs was measured using Aquamax Coulometric Karl Fischer Titrator.

2.3 Solubility procedures

The solubility procedure was conducted for several mixtures of salts and ILs. Table 2 summarizes the salt/IL combinations used for salt solubility test.

Table 2. All salt/IL combinations used for the solubility measurements.

Sample	IL	Salt
1	IL 14	$CaCl_2$
2	IL 14	$MgCl_2$
3	IL 14	NaCl
4	IL 15	$CaCl_2$
5	IL 15	$MgCl_2$
6	IL 15	NaCl

The samples presented in Table 2 consisted of roughly 3g of IL and 0.2g of salt. Each of the samples tabulated in Table 2 was mixed at a set of different temperatures (25, 45, 60, 80 and 100 °C) using the Eppendorf Thermomixer® C (5 mL) at a speed of 600 rpm for 16 to 24 hours and then allowed to settle down. At each temperature, a sample of 0.1 g of the IL mixture was taken and placed in a 50 mL polypropylene centrifuge tube. Each sample was diluted with 50 mL of deionized water in order to accommodate the ICP equipment specification. The diluted samples were analyzed using ICP.

3 Results and discussion

The experimental results are based upon two main experimental procedures; first, solubility of salts in Hydrophilic ILs at different temperatures, and second, the solubility of salts in Hydrophobic ILs at different temperatures. Salts solubility was tested individually in each IL. The solubility of salts in the tested ILs at different temperatures is shown in Tables 3, 4 and 5. Also, Karl Fischer Titration was used to determine the water content of pure ionic liquids. The water content data obtained from the Karl Fischer Titration is reported in Table 5. The water content of ILs is very important since it can affect the physiochemical properties of ILs. As mentioned earlier, the first step for the research analysis is to measure the solubility of the three salts in ILs and keep the results as a reference. The solubility was measured for each salt separately in each of the listed ILs in Table 1 at different temperatures.

Firstly, the solubility of $MgCl_2$ was measured in the ILs at different temperatures. The results are listed in Table 3 and shown in Figures 1 and 2.

Table 3. Solubility results for $MgCl_2$ at different temperatures.

$MgCl_2$					
IL	Temperature (°C)				
	25	45	60	80	100
	Concentration (ppm)				
15	220.55	90.95	51.85	169.05	149.63
14	3,809	8,275	8,478	9,540	15,115

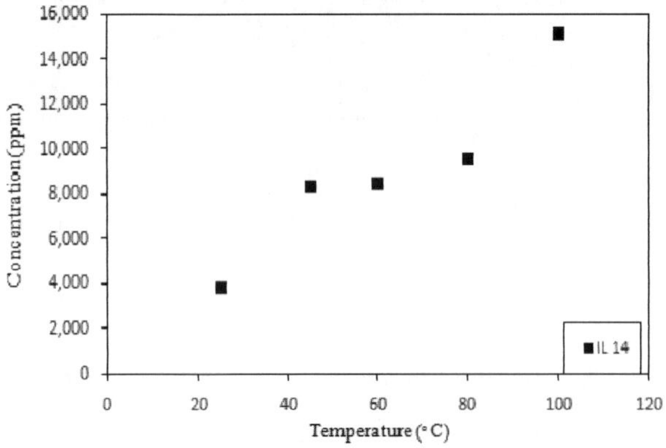

Figure 1. $MgCl_2$ solubility in IL14 at different temperatures.

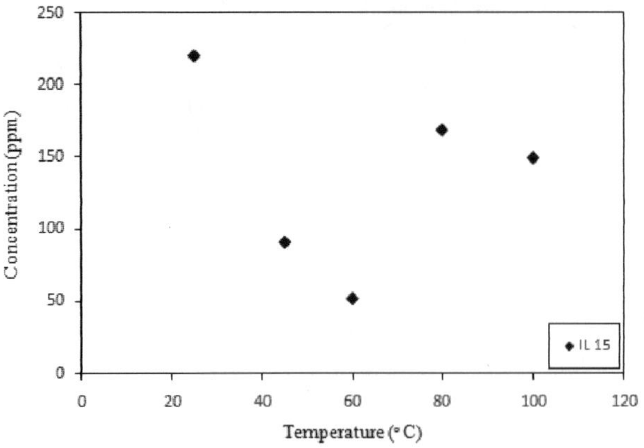

Figure 2. $MgCl_2$ solubility in IL15 at different temperatures.

IL14 has shown the best solubility results of $MgCl_2$ at all temperatures with the maximum concentration of 15,115 ppm at 100°C. The results have shown a trend of increasing solubility with increasing temperature, but this was not followed by IL15. The solubility of $MgCl_2$ decreased from 25°C to 60°C and then increased.

Then, the solubility of NaCl was examined at different temperatures and listed the results in Table 4 and plotted in Figure 3 and Figure 4. Again, IL14 showed the best solubility amongst other ILs with the highest solubility of 6,639 ppm at 60°C and remained approximately constant with increasing temperature. Additionally, the hydrophobic IL15 showed the lowest results that remained constant throughout the temperatures. Clearly, the solubility of $MgCl_2$ in the tested ILs is much higher than NaCl.

Table 4. Solubility results for NaCl at different temperatures.

	NaCl				
IL	Temperature (°C)				
	25	45	60	80	100
	Concentration (ppm)				
15	85.25	84.34	85.07	82.67	84.03
14	1,045	5,845	6,639	6,504	6,540

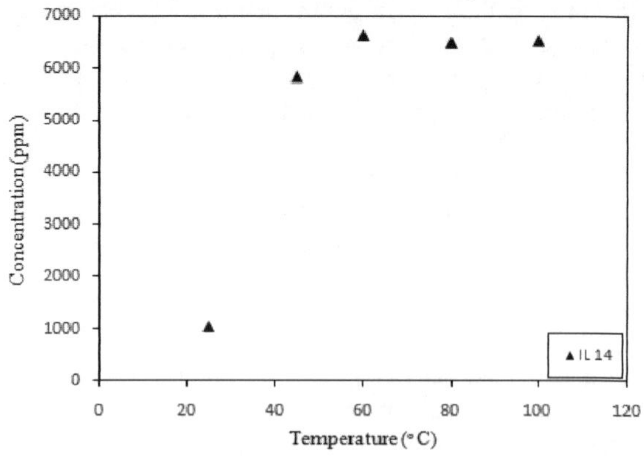

Figure 3. NaCl solubility in IL14 at different temperatures.

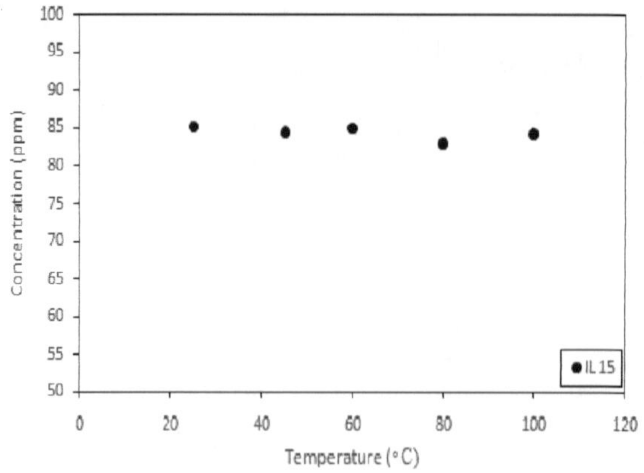

Figure 4. NaCl solubility in IL15 at different temperatures.

As per of $CaCl_2$, the solubility results are presented in Table 5. The highest solubility was 1,393 ppm at 25°C in IL14. It was suspected that a chemical reaction took place in which the ions were exchanged between the salt and IL; cation (Calcium) from the salt transferred to the ionic liquid and combined with the IL anion (dimethyl phosphate) and the cation from the ionic liquid transferred to the salt. Further analysis was performed using FTIR which confirmed that a chemical reaction did indeed take place. We couldn't further observe the solubility change with temperature as it solidified at 45°C. As shown in Figure 5, for IL15, an emulsion formed through 45°C to 80°C which prevented measuring the solubility, but the concentration decreased between 25°C and 100°C which contradicts with the trend showed by the previous salts.

Table 5. Solubility results for $CaCl_2$ at different temperatures.

	$CaCl_2$				
IL	Temperature (°C)				
	25	45	60	80	100
	Concentration (ppm)				
15	570.10	Emulsion	Emulsion	Emulsion	224.60
14	1,393	Emulsion	-	-	-

 https://doi.org/10.1051/matecconf/201818701003

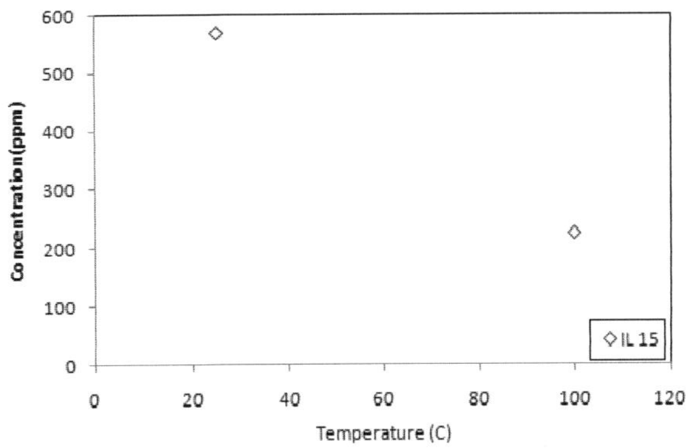

Figure 5. $CaCl_2$ solubility in IL15 at different temperatures.

An experiment was conducted using the Karl Fischer Titrator to measure the water content in pure IL. Table 6 shows the results for the percentages of water content in pure IL. Water content helps to determine whether ILs can be recycled and reused for further saline water extraction processes as we consider the efficient ionic liquid that has the least water uptake; thereby, we reduce the amount of resources needed and the cost of the overall process.

Table 6. Water content of ILs using Karl Fischer Titrator.

IL	% of Water content in Pure I.L.[b]
15	0.0928%

[b]The K-F titration test was done on the pure hydrophobic ILs.

4 Conclusion

In this work, the potential of using ionic liquids as a part of the pre-treatment process for seawater desalination was investigated. The analysis was initiated by measuring the solubility of sodium chloride, magnesium chloride, and calcium chloride in the selected ionic liquids at different temperatures. The solubility of all three salts increased with increasing temperature. IL14 has shown the best solubility results with $MgCl_2$ salt with the maximum concentration of 15,115ppm at 100°C, and NaCl with 6,639 ppm concentration at 60°C, and CaCl2 with 1,393 ppm at 25°C. The results showed that the solubility of $MgCl_2$ in the tested ILs is much higher than that of NaCl and $CaCl_2$. It was found that IL/$CaCl_2$ mixture solidified at temperatures higher than 25°C.

In conclusion, the overall results of the research work show a great potential of using ILs in the pre-treatment process as it reduces the fouling effect on the membranes by primarily extracting the inorganic salts. Also, considering using an efficient ionic liquid with lowest water uptake that can be recycled and reused again for repeated extraction processes.

References

1. R. Dashtpour and S. Al-Zubaidy, Intern. J. of Envi. Sci. and Develop., **3** (2012)
2. Walden, P. Bull Acad. Imper. Sci., 1291. (1929)
3. C. L Hussey Adv. Molten Salt Chem., **5**, 185. (1983)
4. T. Welten, Chem. Rev., 2071. (1999).
5. A. Pereiro, J. Araújo, F. Oliveira, J. Esperança, J. Canongia Lopes, I. Marrucho et al. , J. of Chem. Thermo., **55,** pp:29-36. (2012)
6. F. GharehBagh, F. Mjalli, M. Hashim, M. Hadj-Kali and I. AlNashef, Ind. & Eng. Chem. Res., **52**, pp.11488-11493. (2013).
7. C. Janssen, M. Kroon, S. Metz, J. van Spronsen and G. Witkamp, J. of Chem. & Eng. Data, **55(9),** pp. 3391-3394. (2010).

Effect of Roasting and Kneading on Antioxidant Activity and Consumer Acceptance towards Asiatic Pennywort Tea

Rungnattakan Ploenkutham[1], Preeyapa Sripromma[1], Suksan Amornraksa[2], Patchanee Yasurin[1] and Aussama Soontrunnarudrungsri[1]

[1] *Department of Food Biotechnology, Faculty of Biotechnology, Assumption University, Bangkok, Thailand*
[2]*King Mongkut's University of Technology North Bangkok, Bangkok Thailand*

Abstract. Nowadays, population have more age and longer life than ancient people which the World Health Organization (WHO) reported that the proportion of the global population will become 'Aging Society' during healthier lifestyles, new advance technologies especially in medical manufacture. This study was aimed to study attitude, behavior and opinion of elderly people towards herbal tea products and to optimize preparation, procedure and process of Asiatic Pennywort tea. According of the study, there are 100% of Thai people ever consume herbs and herbal tea, they consume as beverage. Half of consumer surveys are no chronic health conditions. Safety of product is the most factors that effect to purchasing decision. The consumers are agree (73%) in nutrition value that presented on label as it's clamed on the package. Result from this study shown that roasted and kneading process is highest mean score from consumer's preference 7.07±1.51 ($P<0.05$) that this method is significantly different in total phenolic content and Ferric reducing antioxidant potential assay but not in DPPH radical scavenging which is highest mean of second significantly level.

1 Introduction

There are more than 7 billion living humans on earth and the rate of population in each country is increasing that is determined by births and deaths. Asia is the most populous region in the world or 60% of world population which China is the highest population in the world and follows with India. Rank of Thailand is 19 which have 68 million humans. United Nations Population Fund (UNFPA) reported that the world population has less amount of children population that means elderly population are increasing. The world society will change to Ageing Society. Thailand has chance to be Aging Society because the rate of elderly population is 5.7% in 1984 and grow up to 9.6% in 2003 which can expect that Thailand will has elderly population to 14.7% in 2019. Elderly definition is an old person who has age more than 60 years old. When entering old age, your senses (taste, smell, touch, vision, and hearing) become less acute and have some trouble in daily life. Herbs and spices have used as a tradition flavor enhancement for each characteristics and their medicinal properties. Herbs and spices are small plant that does not produce woody persistent tissues and has medicinal properties. Herbs and spices can use different parts of plant through processes such as roasting, smoked, fermented or fresh which consume herbs and spices in small amounts can help in bioactivities in human body which can provide health benefits for consumers [1-4]. Thai people have consumed herbal plant as medicines. Herbal plant can use in many ways. Tea is dried leaves of *Camellia sinensis* plant that usually consume by infuse in hot water. *Camellia sinensis* plant contains all kinds of great properties, including vitamins (E, C), caffeine and antioxidants. There are several types of tea that made from *Camellia sinensis* plant. In each types of tea made from different parts such as green tea, black tea, white tea and oolong etc. and different processes such as roasted, air dried and fermentation to create the specific flavors. Herbal tea will be separate from tea. Herbal tea can call "tisanes" (tea zahn) that is commonly used plants from herbs or spices but normal tea use *Camellia sinensis* plant. Herbal tea is the beverage that can get from using parts of herbs or spices then pass though air dried or fermentation process by pass through soaking, brewing or boiling process as normal tea. There are several herbs that are popular in Thailand and contain several benefits with human health. However, there are the most popular one is 'Asiatic Pennywort' or *Centella asiatica* or Gotu kola. Thailand has 4 herbs popular products which is Gotu kola (Asiatic Pennywort), Tumeric, Phlai (Zingiber cassumunar) and Black galingale (Kaempferia parviflora Wallich. ex Baker.) that is Thailand's Signature of herbal products. Asiatic Pennywort is important herbal medicinal plant used for many benefits [5]. Ancient people was use in treating all disease for long time such as wound healing and eczema.[6-7] Asiatic pennywort contains biological activities compounds for human health, during it contain antioxidant [8-11] that can help in anti-inflammatory[12] and memory enhancing property[10-11]. The main active constituents of *Centella asiatica* are triterpenes [13], polyphenols[14] and triterpenes [15] such as asiatic acid, asiaticoside, medecasic acid and madecassoside that resistant with antioxidation reaction[16]. Asiatic acid in the triterpenes component responsible for stimulating collagen synthesis of human fibroblasts[17]. Free radicals have been played an important role in ageing capable of damaging many cellular components[18]. Antioxidant will protect oxidative stress reaction. Reactive species (ROS) is low stable molecule, it need an electron to make itself more stable. So, reactive species will bring electron form other cells as a cycle that effect to mitochondria membrane and cell will damage. Protection against free radicals can be enhanced by intake of dietary antioxidants which may be major importance in disease prevention. Therefore, the aim of this project was to study the effect of roasting and kneading on antioxidant activity and consumer acceptance of Asiatic Pennywort tea.

2 Materials and methods

2.1 Study attitude, behavior and opinion of elderly people towards tea products

There were 100 people participated in this survey. Questionnaire was developed as a tool to gather the information which it was divided into 6 parts which were Part 1: Consumer's behavior about herbs and tea, Part 2: Consumer's attitude and purchasing behavior, Part 3: The factors influence purchasing decision. this part was asked based on their important factors by rating level of importance which used 1 referred to not at all importance and five referred to extremely importance, Part 4: Food consumption behavior based on their knowledge which ask based on their thinking with ability and expectation from products, Part 5: General Information included questions regarding demographic and socioeconomic position of participants such as age, gender, education, house members, income, occupation and chronic health conditions.

2.2 To optimize preparation, procedure and process of Asiatic Pennywort tea

2.2.1 Procedure

Asiatic Pennywort was purchased from local markets in Bangkok, Thailand. The Asiatic Pennywort was graded that the leave should not have yellow color, bite from insects and free from diseases and pesticides. Fresh Asiatic Pennywort was wash with tap water or rinse before use to clean any soil and dust. The Asiatic Pennywort was packed and kept in refrigerator. There are four formulates with different methods: Kneading & roasted, Kneading & not roasted, Not kneading & roasted and Not kneading & not roasted. Firstly, the Asiatic Pennywort was separated the leaves out from the stalk. The leave was steamed 90 °C for 1 minutes. Cool down at room temperature. After that for kneading process, kneading the leaves by using the hands for 5 minutes and left the leaves sample at room temperature for 30 minutes. For roasted process, after cool down the leaves sample was roasted in the pan at 45 °C for 25 minutes. After that, the leaves were unfolded before put in hot air oven at 80 °C for 2 hours.

2.2.2 Sensory testing and sample preparation

There were 30 elderly people with age between 50 – 75 years old participated in the test whose can be male or female. 2 grams of Asiatic pennywort dried leave was put in tea bag. Brew tea with 150 ml of 90 °C water. The herbal tea bag was soaked for 5 minutes. After that, the herbal tea bag was shaken up and down for 10 times and serve at 65°C - 70°C.

2.2.3 Chemical analysis

The Asiatic pennywort tea was determined amount of antioxidant activities. The basic methods were used. The modified Folin–Ciocalteu method[19], DPPH radical scavenging[20] and Ferric reducing antioxidant potential assay[21] were used for evaluating antioxidant activity.

2.2.3.1 Antioxidant activity by determine total phenolic content

The tea sample The tea sample 20 µl of 10 mg/ml was mixed with 1.58 ml of distilled water and 100 µl Folin–Ciocalteu phenol reagent [22]. The samples were stood for 8 minutes 30 seconds at room temperature. Then, add 300 µl saturated sodium carbonate solution and incubated in dark place at room temperature for 30 minutes. After that, the test samples will be observed the optical density (OD) at 765 nm. The data was calculated as microgram garlic acid equivalent (µgGAE/ml). The experiment was done in triplicate[23].

2.2.3.2 Antioxidant activity by DPPH radical scavenging activity

DPPH radical scavenging activity assay[20]. This assay uses to show free radical scavenging activity between DPPH solvent (2, 2-diphenyl -1-picrylhydrazyl) and an antioxidant. Tea sample 2 µl was mixed with 2 ml of methanol. Then, DPPH was mixed with 100 ml of methanol. After that, volume of DPPH reagent was varied with the volume of tea sample. The varied sample was leaved in the dark place for 30 minutes. The test samples was measured the optical density (OD) at 517 nm. The result will be expressed as percentage reduction of DPPH. All measurements should analyze in triplication and three replications.

$$\%\ \text{reduction of DPPH} = 100\left(\frac{A_0 - A_C}{A_0}\right) \quad (1)$$

Where A_0 is the initial absorbance and A_C is the value for add sample concentration.

2.2.3.3 Antioxidant activity by Ferric Reducing Antioxidant Potential Assay (FRAP)

The tea sample was used for 2 µl mixed with FRAP reagent 1,000 µl, mix well. The mixture was kept in the dark place at room temperature for 30 minutes. After that, the mixture solution was measured the optical density (OD) at 593 nm. The value will be calculated as mmol Fe^{2+}/mg of sample. All measurements should analyze in triplication and three replications. FRAP reagent is mixture between sodium acetate buffer pH 3.6, Iron chloride and 2,4,6 tripyridyl-s-triazine.

2.2.4 Statistical analysis

Data of survey information were processes using SAS version 9.4 (Copyright © [2017] SAS Institute Inc., Cary, NC, USA.), Randomized Complete Block Design (RCBD) was applied as experimental design. The data in this experiment was analyzed by using ANOVA (analysis of variance) and Fisher's LSD test($p<0.05$) for multiple comparison. Sensory testing was used about liking score with 9 point hedonic scale.

3 Result and discussion

3.1 Study attitude, behavior and opinion of elderly people towards tea products

3.1.1 Demographic information

There are 100% of Thai people and nationality who ever consumed herbs and herbal teas. There were male 35% and female 65% participated in this survey with age of 51-55 years old(48%), 56-60 years old(31%), and 61 - 65 years old (17%). Most of the elderly stayed with family that had 4-5 people(46%), 2-3 people (33%), more than 5 people (11%) and stay alone(10%). They obtained in Bachelor degree (37%), Lower than high school (20%), Master degree or higher(19%), High school(13%) and Vocational Certificate or High Vocational Certificate(11%). The income that they earned in a month is more than 50,000 baht(25%), 20,001 – 30,000 baht(21%), and 10,001 - 20,000 baht(19%).Most of them are no health conditions or chronic health(39%), Hypertension(22%), Dyslipidemia (15%), Diabetes mellitus(10%), Obesity and overweight (6%).

3.1.2 Consumer behavior

They consumed Thai herbs (66%) and consumed as beverage (42%), as medicines (31%) and eat as food in meals (27%). The highest herbal tea that consumers had ever consumed is Chrysanthemum tea (16%), Ginger tea (14%) and Lemon glass tea (13%). They like to consume They tea from tea bag (34%) and consume only 1 time in a month (23%) which no specific the time to consume herbal tea (45%), in morning (26%) and afternoon (21%). Consumers purchase herbal tea at hypermarket the most (37%), and in each time will purchase more than 75 baht (41%) for herbal teas. The consumers agreed (73%) in nutrition value that presented on label as it's clamed on the package which they think that their body will absorb the nutrients from herbal teas only 50 percent (39%), 25 percent (33%) and 75 percent (24%). The reason of consumers consume herbal tea is contain medical benefit (28%), make them feel fresh (23%) and relaxing (18%).

Table 1. The important factor effect on purchasing decision.

Factors	Mean ± SD
Safety	4.5 ± 0.9^{a}
Certification	4.3 ± 1.0^{b}
Nutrition	4.3 ± 1.0^{b}
Taste	4.1 ± 0.9^{bc}
Medical properties	4.0 ± 1.0^{c}
Organic	3.7 ± 1.2^{d}
Source	3.6 ± 1.2^{de}
Package	3.4 ± 1.2^{ef}
NET	3.3 ± 1.1^{ef}
Brands	3.2 ± 1.1^{fg}
Trend	3.1 ± 1.2^{g}

There are significant different between factors effect on purchasing decision ($p<0.05$). The most important factor is a safety; it means that they like to safe in consume the tea. Next factors are certification, nutrition, and taste. It means that consumer want to consume good products which should has approved from FDA or other warranties due to their age. The taste is also significantly important in the same level as medical properties. Taste relate to perception of consumer that should be good and more benefits together. Moreover, they are not concern in organic product or source, the reason might be the properties contain in herbs are mostly the same benefits. Package, Net weight and Brand of herbal tea products also not affect to purchasing decision. Paper tissue has used for package of tea both inside or outside for tea infusion. Plastic bottle has used as package of tea bottle that easy to consume than tea infusion or when you want to consume tea immediately. Volume or weight is not affected to their decision. Elderly people is not care much in the volume because they are more concern in taste than others factors. From survey found that they are consume only 1 time in a month. Brand and trend are lowest factors for elderly people; it means that famous products, social products or popular products are not effect to their purchasing decision. Nowadays, several was produced and lunched by small group of villagers or they can make it by themselves.

Table 2: Liking score of Asiatic Pennywort tea from 30 consumers in 10 attributes.

Process	Overall liking*	Color	Greeny aroma*	Roasted aroma	Sweet
Roasted & Knead	7.1 ± 1.5^a	7.4 ± 1.3^a	7.1 ± 1.1^a	6.6 ± 1.5^a	6.8 ± 1.2^a
Roast & no kneed	6.9 ± 1.3^a	7.1 ± 1.0^{ab}	6.6 ± 1.5^{ab}	6.3 ± 1.7^a	6.7 ± 1.4^a
No roast & kneed	6.3 ± 1.5^b	7.1 ± 1.3^{ab}	6.3 ± 1.6^b	6.3 ± 1.1^a	6.4 ± 1.3^a
No Roast & no kneed	6.2 ± 1.4^b	6.6 ± 1.4^b	6.2 ± 1.5^b	6.1 ± 1.4^a	6.2 ± 1.7^a
Process	Bitter	Greeny Flavor	Roasted Flavor	Greeny Aftertaste *	Astringent
Roast & kneed	6.5 ± 1.3^{ab}	6.9 ± 1.6^a	6.7 ± 1.4^a	6.9 ± 1.6^a	7.0 ± 1.4^a
Roast & no kneed	6.8 ± 1.3^a	6.8 ± 1.4^a	6.6 ± 1.3^a	6.7 ± 1.2^{ab}	6.7 ± 1.3^{ab}
No roast & kneed	6.1 ± 1.2^b	6.4 ± 1.5^a	6.2 ± 1.3^a	6.1 ± 1.5^{bc}	6.5 ± 1.6^{ab}
No Roast & no kneed	6.3 ± 1.5^{ab}	6.5 ± 1.5^a	6.2 ± 1.6^a	6.0 ± 1.6^c	6.1 ± 1.9^b

Table 3 : Result of Total phenolic content, Ferric reducing antioxidant potential and DPPH radical scavenging of Asiatic Pennywort tea from different processes.

Process	Total phenolic content (µg GAE/mg dried weight)	DPPH (% reduction)	FRAP (mmol Fe^{2+}/mg dried weight)
Roast & kneed	62.7 ± 32.8^a	0.2 ± 0.0^b	1.6 ± 0.1^a
Roast & no kneed	28.6 ± 4.4^b	0.1 ± 0.0^b	1.7 ± 0.4^a
No roast & kneed	54.0 ± 14.7^{ab}	0.3 ± 0.1^a	1.4 ± 0.6^a
No Roast & no kneed	41.5 ± 10.7^{ab}	0.2 ± 0.0^b	1.9 ± 0.2^a

3.2 To optimize preparation, procedure and process of Asiatic Pennywort tea

3.2.1 Sensory analysis

All attributes were selected from characteristic of sample. There were 30 consumers that participating in the sensory test. There are significant different in overall liking, greeny aroma and greeny aftertaste after taste in different processes of tea ($p<0.05$). The significantly most preferred for all attributes is Roast & kneed process and Roast & no kneed process, it means that most of consumers are like to consumer roasted processes. Two processes are similar in roasted aroma, sweetness, greeny flavor and roasted flavor. But others all attributes were second level that is color, greeny aroma, greeny aftertaste and astringent of Roast & no kneed process. The bitterness of Roast & kneed process is the second level. Moreover, greeny aftertaste has a lowest mean score from consumers which sample No roast & kneed process and No roast & no kneed process. From result, roasted greeny aroma and greeny aftertaste, both attributes has lowest mean score in the same process that is not roasted. Roasted also help in bitter taste process can help to reduce greeny attributes. Focusing on (vellarine compound) and astringent (tannin compound) which heating can reduce the level of tannin in winged bean. Roasted also reduce greeny aroma that occur in the tea products [24-25].

3.2.2 Chemical analysis

Herbs are rich in phenolic compounds which are active compound and act as antioxidant properties. Asiatic Pennywort also has active compounds such as triterpenes [13] and polyphenols [14] which contains high total phenolic contents. Total phenolic content was determined with standard garlic acid in term of µg GAE/ mg dried weight. The highest antioxidant activity represented by amount phenolic compound that is 62.7 ± 32.8 µg GAE/mg dried weight which use Roasted & kneading process. This process is most significant different but not significant different from another 2 processes that is Not roasted & kneading and Not roasted & not kneading ($p<0.05$). Thermal process is a big influence in active compound which depend on their magnitude, duration and different heating methods. Thermal process lend to degradation of phenolic compounds. For example, roasting barley at 280 °C for 20 seconds can decrease 8% of phenol content[26]. For DPPH assay is significantly different in all processes that the highest mean is Not roasted & kneading process 0.3 ± 0.1 % DPPH radical scavenging ($P < 0.05$). Using convection oven drying with spearmint can decrease antioxidant activities antioxidant activities approximately 60% which heating go to inactivate enzyme. [27]. Kneading was used to macerate the cells of leaf which can activate enzymes (mixing of chemical with enzyme) inside the leaf and can be easily being extracted during brewing.[28] For Ferric reducing antioxidant potential assay, there are not significantly different in all processes which the highest mean score is Not roasted & not kneading process 1.9 ± 0.2 mmol Fe^{2+}/mg ($P < 0.05$).

4 Conclusion

The obtained result in this project was used to study habit, attitude and decision of elderly people. The process that will be selected to produce the Asiatic Pennywort tea depend on what is the most concern that if we concern at highest in both testing from sensory analysis from consumer and active compound, Roasted & kneading process will be selected to produce the Asiatic Pennywort tea. This process is get most significant different in mean scores for sensory, total phenolic content and ferric reducing antioxidant except DPPH radical scavenging has highest mean score from second significant.

Acknowledge

The authors would like to thank National Research Council of Thailand (NRCT) for the financial support of this research. This work was financially supported by King Mongkut's University of Technology, North Bangkok (Research University Grant No. KMUTNB-61-GOV-01-55).

Reference

1. Ferrucci et al. Measurement of spices and seasonings in India: Opportunities for cancer epidemiology and prevention, *Asian Pacific Journal of Cancer Prevention*, 1621-1629, (2010)
2. Kaefar M., The role of herbs and spices in cancer prevention *Nutritional Science Research Group, National Cancer Institute*, Rockville, MD 20892, USA (2008)
3. Johanna L., Spicing up a vegetarian diet: Chemopreventive effects of phytochemicals, *The American Journal of Clinical Nutrition*, **78** ,579-583 (2003)
4. Tierano L., A reason to season: The therapeutic benefits of spices and culinary herbs, *The Journal of Diet and Nutrition*, **2**, 446-449 (2006)
5. Jacainda J. and Ian D., Pentacyclic triterpenoids from the medicinal herb: Centella asiatica (L.) Urban. Molecules, **14**, 3922-3941 (2009)
6. Brinkhaus, B., Lindner, M., Schuppan, D. and Hahn, E. G., Chemical, pharmacological and clinical profile of the East Asian medical plant Centella asiatica. Phytomedicine, **7**, 427-428 (2000)
7. Kimura, Y., Sumiyoshi, M., Samukawa, K., Satake N. and Sakanaka M.; Facilitating action of asiaticoside at low doses on burn wound repair and its mechanism. *European Journal of Pharmacology* 584: 415-421. (2008)
8. Hamid, A. A, Md Shah, Z., Muse, R. and Mohamed, S. Characterisation of antioxidative of Centella asiatica (L) Urban. *Food Chemistry* **77**: 465-469 (2002)
9. Pittella, F., Dutra, R. C., Junior, D. D., Lopes, M. T. and Barbosa N., Antioxidant and cytotoxic activites of Centella asiatica (L) Urb. *International Journal of Molecular Science* **10**, 3713-3721, (2009).
10. Ullah, M. O., Sultana, S., Haque, A. and Tasmin, S. Antimicrobial, cytotoxic and antioxidant of Centella asiatica. *European Journal of Scientific Research* , 260-264 (2009).
11. Vimala, S., Adenan, M. I., Ahmad, A. R. and Shahdan, R. Nature's choice to wellness: antioxidant vegetables/ulam. Siri Alam dan Rimba, 90-92 (2003).
12. Duke, J. A. (2001). *Handbook of Medicinal Herbs*. 1st edition. New York:
13. Rattanakom, S.; Yasurin, P. Antibacterial activity, antioxidant activity and chemical profiling of Centella asiatica under different extraction solvents. *Orient. J. Chem* **31**, page 2453-2459 (2015)
14. Zainol A, Hamid S, Yusof R, Muse. Antioxidative activity and total phenolic compounds of leaf, root and petiole of four accessions of Centella asiatica(L.) Urban. *Food Chemistry*. **81**, 575-581(2003)
15. Inamdar, P.K.; Yeole, R. D.; Ghogare, A. B.; de Souza, N. J. Determination of Biologically Active Constituents in *Cente-lla Asiatica. J. Chromatogr. A.*, 742, 127-130 (1996)
16. Ling, A.P.K.; Marziah, M.; Tan, S.E. Triterpenoids distribution in whole plant and callus cultures of *Centella asiatica* accessions. *In Proceeding of the 16th National Seminar on Natural Products,* 165-168 (2000)
17. Maquart, F.X.; Bellon, G.; Gillery, P.; Wegrosky, Y.; Borel, J.P. Stimulation of collagen synthesis in fibroblast cultures by a triterpene extracted from Centella asiatica. Connect. Tissue Res, **24**, 107-120 (1990)
18. Gulcin, I., Oktay, M., Kirecci, E. and Kufrevioglu, O.I. Screening of antioxidant and antimicrobial activities of anise (Pimpinella anisum L.) seed extracts. *Food Chemistry* **83**: 371-382 (2003)
19. Ragazzi, E.; Veronese, G. Quantitative analysis of phenolic compound after thin-layer chromatographic separation. *J. Chromatogr. A.* (1973)
20. Brand-Williams, W.; Cuvelier, ME.; Berset, C. Use of a free radical method to evaluate antioxidant activity. *LWT-Food Science and Technology.* (1995)
21. Benzie, IF.; Strain, JJ. Ferric reducing/ antioxidant power assay: Direct measure of total antioxidant activity of biological fluids and modified version for simultaneous measurement of total antioxidant power and ascorbic acid concentration. *Methods Enzymol*. (1999)
22. Samtram S. Effect of heat treatments of antinutritional factors and quality of protein in winged bean. *Journal of the science of food and agriculture,* **39,** 267-275 (1987)

23. Kesornbuakao & Yasuri.; The Development of Centellaasiatica Extract-Loaded BSA Nanoparticles Production to Improve Bioavailability, *Orient. J. Chem.,* Vol. **32**(5), 2425-2434 (2016)
24. Liang, Y., Zhang, L., & Lu, J. A study on chemical estimation of pu-erh tea quality. Journal of the Science of Food and Agriculture, **85**, 381–390 (2005)
25. Zhang, J., Zhang, Z. Z., Zhou, Y. B., Ling, T. J., & Wan. Chinese dark teas: Postfermentation, chemistry and biological activities. *Food Research International* **53**, 600-607 (2013)
26. Sharma, P. & Gujral, H.S. Effect of sand roasting and microwave cooking on antioxidant activity of barley. Food Research International, **44**, 235-240 (2011).
27. Lim Y.Y., Murtijaya J. Antioxidant properties of Phyllanthus amarus extracts as affected by different drying methods. *LWT-Food Science and Technology*, **40**: 1664–1669 (2007)
28. Naheed Z et al,. Effect of Rolling, Fermentation and Drying on the Quality of Black tea. *Sarhad Journal of Agric*, **23**, 577-580 (2007).

Extraction and Characterization of Omega-3 Fatty Acid from Catfish Using Enzymatic Hydrolysis Technique

N.I. Iberahim[1], Z. Hamzah[1], Y.J. Yin[1], and K.S.A. Sohaimi[1]

[1]*Faculty of Engineering Technology, Universiti Malaysia Perlis, Sungai Chuchuh, 02100 Padang Besar, Perlis, Malaysia.*

Abstract. Extraction of Omega-3 fatty acid from Catfish using enzymatic hydrolysis techniques are expected to be more economically possible techniques due to the uses of enzyme with the characteristic of reusable, environmental friendly and less energy required for large-scale production. Malaysian like to consume Catfish in their daily diet. Thus, it is very popular among farmers in Malaysia. Omega-3 polyunsaturated fatty acid have many benefit to human health. Based on research, EPA and DHA can reduce the risk of heart disease by regulate the blood coagulation. The main purpose of this research is to extract the Omega-3 polyunsaturated fatty acid by using enzymatic hydrolysis method. Enzymatic extraction technique by using Alcalase was used to extract the Catfish oil. Then, lipase was used to concentrate the omega-3 polyunsaturated fatty acid. Peroxide value and acid value were used to determine the Catfish oil quality. Optimization using DOE verify that the highest percentage of oil yields at 51°C, 0.91% enzyme concentration and 155 minute incubation time. Finally, there are two types of omega-3 polyunsaturated fatty acid can be found in Catfish oil which is EPA and ALA.

1 Introduction

Catfish is one of the diverse group of ray-finned fish that mostly found in the freshwater of Malaysia. Normally, catfish is grey in colour and scaless. People can identify catfish by their long whiskers-like barbels extend from their mouth. Catfish may consist up to four pairs of barbels. Besides that, catfish also consists of flattened broad head and hollow bony leading ray on their dorsal and pectoral fins. They are usually bottom feeder and they like to prey at night. Catfish are carnivores so they mostly prey small fish, worms, frogs, insects, and newts that present in the river [1].

Freshwater pond culture had been grown rapidly due to the development of aquaculture in Malaysia recently. Catfish is one of the fish that is popular among farmers because its easiness in culture, hardiness, high reproduction rate, and high growth rates. Catfish can survive in varying environment due to their hardiness. Therefore, farmer only needs to use plastic bowl to culture catfish. Catfish also had excellent nutritional value and market value in Malaysia.

Based on the Food and Agriculture Organization (FAO), Malaysia categorized as one of the top fish-consuming countries in Asia which above 40kg/capital/years almost double the average in Thailand and China. Omega-3 fatty acid which is a polyunsaturated fatty acid is one of the significant nutrition in commercial Catfish [2].

Omega-3 fatty acid mostly can be obtained in fish oil. Omega 3 fatty acids is an essential fats that consist of a chain of carbon, hydrogen, and oxygen atoms. It is also categorized as polyunsaturated fatty acid which consists of more than one double bond. Polyunsaturated fatty acid only can be obtained from diet because human body cannot synthesis polyunsaturated fatty acid. Basically, there are two types of omega-3 fatty acids that can be found in fish oil, namely Eicosapentaenoic acid (EPA) and Docosahexaenoic acid (DHA). Based on the research, it is shown that patients with the V hyperlipidemia, fish oil supplements is used as one of the important therapeutic option [3].

There are many benefits of EPA and DHA to human. EPA and DHA both help to regulate blood coagulation and control inflammatory responses. According to the information obtained, EPA used to treat high risk pregnancies, heart disease, depression, diabetes, Alzheimer's disease, high blood pressure, and cystic fibrosis [4]. EPA is modified to become antidepressants to treat depression. Besides that, EPA combined with RNA and L-arginine is used to heal wounds after surgery and reduce the time of recovery of the patient. DHA is important for maintain brain health in which researchers determine that DHA provides brain-boosting benefits in newborn baby and senior citizen. Alzheimer's disease patients consist of low levels of DHA in the neurons of hippocampus. Hippocampus is the disease affected part of the brain. Prasad and co-workers from the University of Kentucky prove that phosphatidyl-ethanolamine contains large amount of DHA that can reduce those brain region that is affected by this diseases [5]. In market, pregnancy woman is also recommended to take milk with high DHA levels to improve brain development of fetus [6].

There are many methods that can be used to extract catfish oil. Enzymatic hydrolysis is one of the methods that uses suitable enzyme and water to extract catfish oil. The benefit of this method is can obtain large amount of oil compared to others method. This is due to the present of enzymes that can catalyze the hydrolysis reaction process

2 Material and method

2.1 Raw materials

The raw material used in this experiment was taken from Pusat Pengumpulan Hasil-Hasil Perikanan Air Tawar, Pejabat Perikanan Air Tawar, Perlis. The Catfish collected at this place was Clarias Gariepinus species. The sample was homogenized and kept in sealed plastics bags in freezer at -20˚C.until further used.

2.2 Experiment design

This experiment operated in two main sections:
- Optimized the optimum condition for extraction of the highest percentage of oil yields.
- Concentrated the omega-3 fatty acid from fish oil.

In the first section, enzymatic hydrolysis was carried out in order to obtain fish oil. Three parameter were screened by using a DOE software Full Factorial Design to ensure that the parameter chosen was significant which can affect the amount of fish oil. The experiment was carried out by using temperature in a range between 35˚C-55˚C. The second parameter is enzyme concentration. 0.5%-2.5% unit of enzyme was used to collect fish oil in this experiment. Lastly, incubation time is one of the important parameter in this experiment. Time of incubation that decide to investigate in this experiment was between 60min-80min. In section 2, a commercial Candida rugose lipase was used to concentrate the amount of Omega-3 fatty acid in the fish oil. The recovery oil yields before and after concentrated by lipase enzyme were then analyzed by FTIR and GC.

2.3 Enzymatic extraction of fish oil

50 g of minced fish sample was transferred into 250 ml of conical flask. Preheated water bath was used to heat the sample at 95˚C for 5 min. Then, 100ml of distilled water was added into the conical flask. The pH of the sample mixture must be maintained at pH 7.0. The pH of the sample mixture was adjusted by adding 3M NaOH. Next, the conical flask was placed into water bath shaker at desired temperature. Enzymatic hydrolysis was settling out when adding Alcalase enzyme into the conical flask. Different load of Alcalase enzyme was used to identify how concentration of enzyme effect on the amount of fish oil production. The enzymatic hydrolysis reaction was carry out in water bath shaker at 120 rpm for desired incubation time. Then, heating the sample at 95˚C for 5 min to denature the structure of enzyme. The top two layer oil and emulsion which then transferred to another centrifuge tube. 2 ml of 0.9% NaCl were added into the centrifuge tube. Another round of centrifugation process was carried out at 3000 rpm for 15 min. The upper layer collected and the lower layer was discard. Then, anhydrous sodium sulfate were added into the top layer have been collected. Then, the oil remain was collected and the anhydrous sodium sulfate was discarded.

2.4 Optimization of enzymatic extraction process

4 g of fish oil were weighed by using weighing balance and transferred into a 50 ml conical flask. The pH of the fish oil sample was maintained at pH 7.0 by added 6 ml of 1M phosphate buffer. Then, 800 unit of lipases was added into the phosphate buffer and transferred into the conical flask with oil sample. The flask was put into water bath shaker at 35˚C for 20 hours. After that, 2 ml of methanol was added to the sample mixture to stop the enzymatic hydrolysis reaction. Fatty acid was released during enzymatic hydrolysis. Thus, desired amount of 0.5N KOH added to neutralize the fatty acid released. The sample mixture was transferred into separator funnel and mixed with 100 ml of hexane and 50 ml of distilled water. The upper layer consists of hexane and acylglycerols. The concentrated oil was kept in universal bottle under refrigerator to undergo further analysis.

2.4 Fatty acid methyl ester (FAME) analysis

The major components of acylglycerols extracted from the previous experiment was fatty acids. Basically, FAME were prepared by base or acid catalyzed esterification. Sample before and after concentrated was used to prepare FAME analysis. 0.1g of lipids from previous experiment were weighted by weighing balance and transferred into a centrifuge tube. Then, esterification reaction was started by adding 1.5 ml of methylene chloride, 0.01% BHT and 3.0 ml of Hilditch reagent into the tube. Heating block was used to place the centrifuge tube for 1 hours. After 1 hours, the tube was cool to room temperature. Next, 1 ml of distilled water and 1.5 ml of hexane added into the tube. The top layer was transferred to the Tube 2 in each repeat experiment. Tube 2 was closed with cap then ready to vortex for 30 s and centrifuge for 2 min. The upper layer of Tube 2 transferred to another new tube which called tube 3. Anhydrous sodium sulphate was added to Tube 3 to removed excess water. The FAME produced will undergoes GC-MS analysis.

2.5 Determination of acid value of oil sample

Two 250 ml conical flask was prepared to carry out this experiment. First, 125 ml of 1:1 toluene-isopropyl alcohol and 2 ml of phenolphthalein indicator was added into the 250ml conical flask. The solution mixture in the first conical flask was neutralized with 0.1N potassium hydroxide solution until a faint but permanent pink colour appear. 0.5-2.5 g of oil sample was weight by using weighing balance and transfer into another empty conical flask. Then, measure 125 ml of neutralized solvent mixture in the first conical flask was measured and pour into the conical flask with oil sample. Then, oil sample and the solvent were mixture carefully. Next, the mixture was titrated with 0.1N KOH solution until pink colour can be observed. The volume of 0.1N KOH solution that used to titrate the mixture solution must be measure. The acid value was calculated by (1):

$$\text{Acid value} = ((A-B) \times N \times 56.1)/ W \quad (1)$$

Where, A= ml of standard alkali used in the titrating the sample, B= ml of standard alkali used in the titrating the blank, N= normality of standard alkali, W= grams of sample

56.1= molecular weight of KOH in grams

2.6 Determination peroxide value of oil sample

Peroxide value of oil sample defined as the amount of peroxide oxygen per 1kg of fat or oil. First, 5 g of oil sample weighted by weighing balance. Then, the oil sample was transferred into 250ml conical flask. After that, measuring cylinder used to measure 30 ml of 3:2 acetic acid-chloroform. Then, 3:2 acetic acid-chloroform transferred into the flask and swirl. 0.5ml potassium iodide solution was added and the solution allowed to stand and put into a shaker for 1min. 30 ml of distilled water was measured and transferred into a flask. Titration carried out with 0.1N sodium thio-sulfate until the mixture solution change to yellow colour. 0.5 ml of starch indicator solution was added into a solution. Titration was continued carry out until the solution turns to colourless. The same condition solution were prepared but without adding oil sample as a blank solution. The peroxide value can be determined by (2):

$$\text{Peroxide value} = ((S-B) \times N \times 1000)/W \quad (2)$$

Where, B= ml of standard potassium thiosulfate used for titration of blank, S= ml of standard potassium thiosulfate used for titration of sample, N= normality of sodium thiosulfate solution 1000= per 1000g of sample, W= gram of sample.

3 Results and discussions

3.1. Parameter and range set in DOE software

Second Full Factorial Design was used to screen the parameter. Three parameter were undergo screening which are temperature, enzyme concentration, and incubation time. The response was taken by the average percentage oil yields obtained. Table 1. represented the result provided by ANOVA analysis. Based on the result of ANOVA, temperature, enzyme concentration, and incubation time have the "Prob > F" value less than 0.0500. Thus, temperature, enzyme concentration, and incubation time is the significant parameter which will affect the oil yields percentage obtained from the experiment.

Table 1. Screening result by anova.

Parameter	Prob > F	Significantly
Temperature	0.0045	Significant
Enzyme Concentration	0.0059	Significant
Incubation Time	0.0107	Signifcant

3.2 RSM-ANOVA analysis

Based on the multiple regression analysis of ANOVA, a quadratic model 3 was generate which is correspond to the oil yields percentage and various process variables.

$$\text{Oil Yields (wt\%)} = 14.67 + 0.55A + 0.40B + 0.31C + 0.41A2 + (6.553e\text{-}0.03)\ B2 + 0.070C2 - 0.24AB + 0.63AC - 0.35BC \quad (3)$$

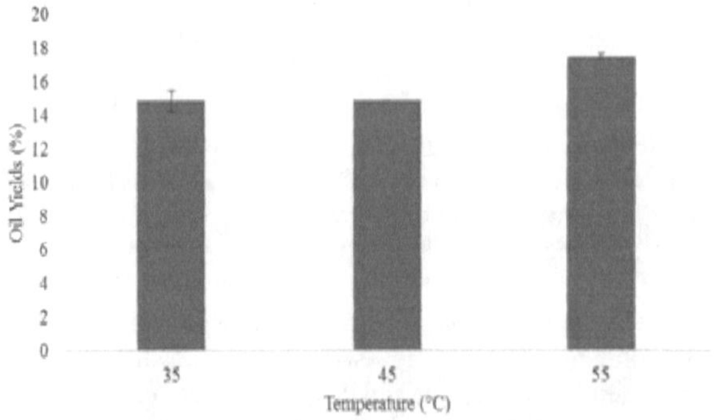

Figure 1. Bar chart of percentage oil yields versus temperature at 120min and 1.5% enzyme concentration.

According to Figure.1, the percentage of oil yields is the highest at 55°C which is 17.5062±0.1606%. However, the percentage of oil yields almost the same at 35°C and 45°C which is 14.8537±0.6163% and 14.8527±0.0073% respectively. This indicated that Alcalase is thermostability, it is not active at low temperature. When the temperature is low, it does not provide enough energy for the Alcalase to bind with the substrate. However, when the temperature increase to 55°C, the percentage of oil yields also increase. This is because, the substrate molecule gain energy and will move faster at high temperature. Hence, increase the temperature will increase the frequency of collision between enzyme active site and substrate [7]. Based on the research of Akika Bura Mohanarangan, Alcalase have the optimum temperature between 50-60 °C [8]. Thus, our result obey this statement because Alcalase have the highest activity at 55°C when compared with others temperature.

3.3 Effect of enzyme concentration

https://doi.org/10.1051/matecconf/201818701005

Enzymatic extraction of Catfish oil start by using the enzyme, Alcalase. The amount of enzyme use in the extraction process strongly influences the percentage of oil yields. The concentration of enzyme added into the enzyme reaction were 0.5%, 1.5%, and 2.5% by 50g of minced Catfish raw materials.

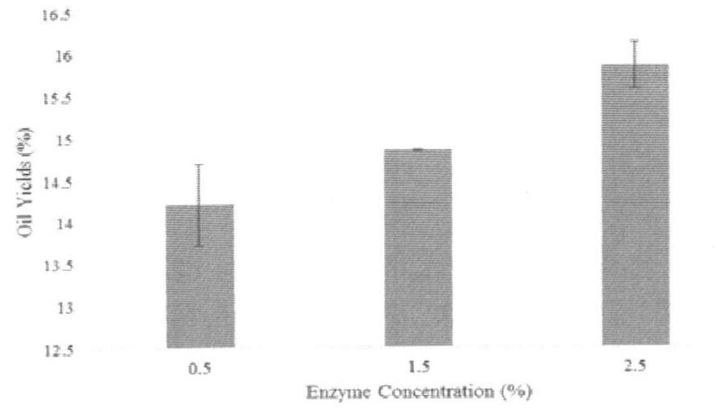

Figure 2. Bar chart of percentage oil yields versus enzyme concentration incubated at 45°C and 120 minutes.

Runs 1, 6, and 7 shows that the percentage of oil yields increase linearly when the enzyme concentration increase in the extraction process. Based on Figure 4.4, when the enzyme concentration use in enzymatic extraction was 2.5% obtained the highest oil yields which is 15.8486±0.274%. When the enzymatic concentration increase, more substrate can be binds to the enzyme. Thus, more free fatty acids will be formed and increase the oil yields percentage. According to research, substrate concentration acts as the limiting factors for the enzymatic extraction reaction. When the substrate concentration use is lower than the enzyme concentration, increase the enzyme concentration will not affect the percentage of oil yields.

3.4 Effect of incubation time

In the optimization process, the range of incubation time were set between 60minute to 180 minutes and observe the effect of incubation time in enzymatic extraction process. Table shows that the F value of incubation time is smaller than temperature and enzyme concentration. This means that incubation time slightly affect the percentage of oil yields.

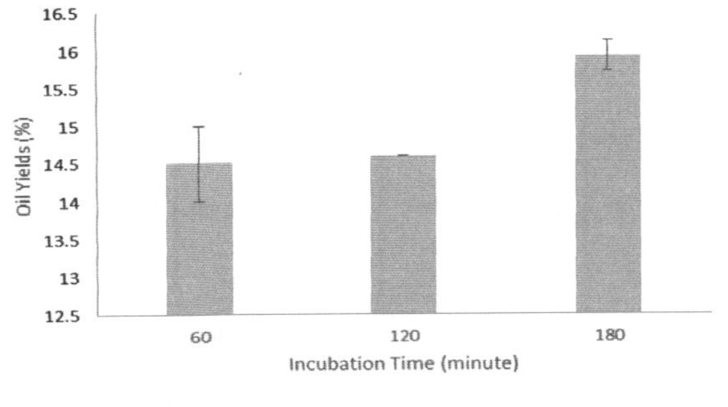

Figure 3. Bar chart of percentage oil yields versus incubation time at 45°C and 1.5% enzyme concentration.

When the incubation time increase from 60 minute to 180 minute, the percentage of oil yields obtained increase from 14.4944±0.5046% to 15.9186±0.2012%. Based on the research, when incubation time increase the rate of hydrolysis also increase. This is because enzyme can react more extensively with substrate and more product will be generated.

3.5 Optimization process parameter

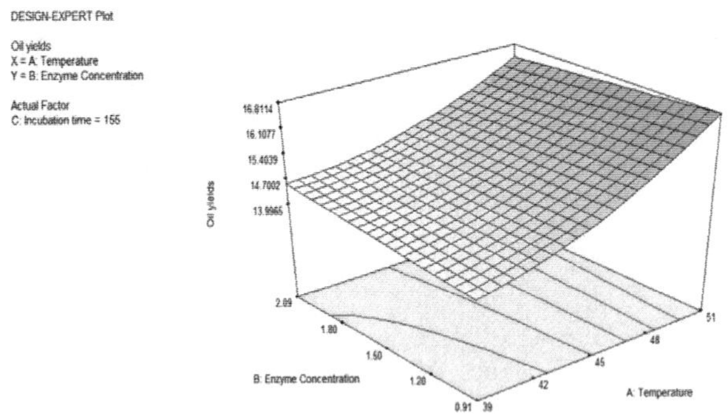

Figure 4: 3D surface of the optimum condition to extract the highest oil yields.

The purpose of optimization process is to identify the optimum condition for each parameter to achieve the highest oil yields percentage in the enzymatic extraction process. The software used to help determine the optimize point is Response Surface Methodology with Central Composite Rotatable Design. The individual process parameter and interaction between parameter was used to analyze and determine the optimum point in enzymatic extraction process. The software provide 10 solution that can achieve high percentage oil yields within the limit. The software predicted that the best solution will produced the highest oil yields percentage 16.8113±0.906% was 0.91%, 51°C, and 155minute. Then, actual experiment was carried out to valid that the optimum point provided by the process is accurate. The actual experiment was run in two times and obtained the average percentage of oil yields. The average percentage oil yields of the actual experiment was 16.5858±0.0694%.

3.6 Functional group FTIR

FTIR is the instrument that used to characterize the functional group of the Catfish oil. Spectrum TwoTM FTIR Spectrometer by Perkin Elmer was used to analysis the Catfish Oil sample before and after concentrated by lipase. FTIR Spectrophotometer by Perkin Elmer was equipped with ATR and DTGS (Deuterated Triglycine Sulfate) detectors to measure the IR spectra of Catfish oil liquid sample. The background of the software was collected before analysis the sample. Micropipette was used to pipette 40ul of sample and placed on the ATR crystal to undergo analysis. The spectra wavelength was set as 4000 cm-1. Perkin Elmer Spectrum Version 10.5.2 was used to obtained database buildings and band assignments.

The Catfish oil before undergoes lipase concentrated consists of carboxylic acids, alkanes, saturated aliphatic,

aliphatic methyl, methyl, alcohol, methylene, and phosphorous functional group [9][10]. However, the Catfish oil after undergoes lipase concentrated consists of alkanes, saturated aliphatic, aldehydes, aliphatic methyl, alcohol, phosphorous, and methylene compound. The knowledge from chemistry describes that fish oil is in the form of triglycerides. Triglycerides is the compound that made up of three ester functional group. In the FTIR spectra shows peaks of 1743.00cm and 1743.08cm which represent the present of ester group in the sample. Thus, in the sample consists of triglycerides compound. Besides that, FTIR spectra peaks 1117.49cm, 1096.71cm, and 1117.48cm indicated the present of glycerol in both sample.

When triglycerides undergoes hydrolysis will produce glycerol and three free fatty acids chain. Moreover, the peaks of 3007.37cm and 1235.95cm represent carboxylic group in the free fatty acids of the sample. Fatty acid chain in the triglycerides are made up of long chain of hydrocarbon. Mostly, they consists of even number of carbon atoms which from 4 to 36 carbon atoms [11]. Aliphatic methyl functional group determine by 1463.69cm and 1464.76 peaks and the methylene rocking vibration 720.10cm and 721.65cm valid the present of long and linear aliphatic hydrocarbon chain in the Catfish oil [12]. Thus, there will be present of methyl group in the fatty acid chain. If there is the single bond between the carbon atoms is called saturated fatty acids. On the others hand, there is the double bond between some of the carbon atoms is called unsaturated fatty acids.

Based on the spectrum diagram we obtained from FTIR analysis, 2921.65cm and 2922.05cm peaks represent alkenes functional so determine there is the unsaturated fatty acid in the Catfish oil. The alkanes group was used to prove the present of saturated fatty acid in the Catfish oil. Besides that, 1159.13cm and 1163.54 peaks determine the present of phosphorous compound [10]. This analysis valid that the Catfish oil consists of phospholipid because there is the present of phosphate group in the fatty acid chain. The main function of lipase was used to concentrate the polyunsaturated fatty acids such as EPA and DHA. Lipase will used to break the ester bond between glycerol and monounsaturated fatty acid and saturated fatty acid. The peaks of 1743.00cm and 1744.08cm is in the range that indicated the present of triglycerides ester bond. The sample before concentrate by lipase 1743cm is higher than 1744.08cm. This is because most of the triglycerides ester bond hydrolyses by lipase to release free fatty acids chain [14].

3.7 FAME analysis

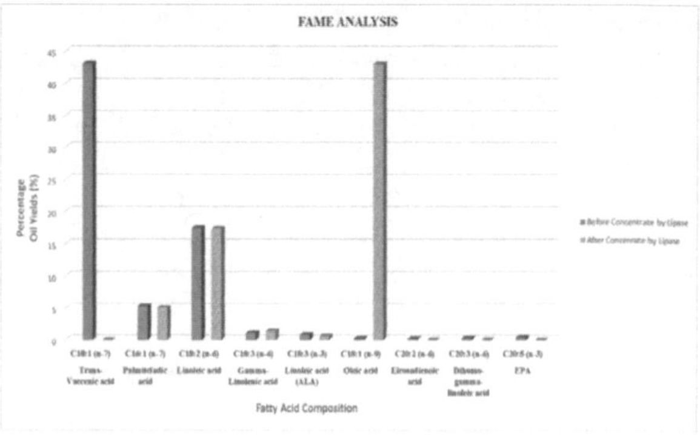

Figure 5. Fatty acid composition of catfish oil.

The Catfish oil that in the triglycerides structure undergoes saponification to produce glycerol and free fatty acids. Then the free fatty acids will undergoes methylation with methanol and derivatization in the methyl ester form and become fatty acid methyl ester (FAME). Gas Chromatography only can separate the fatty acid composition in the methyl ester form. This equipment separate the fatty acid based on degree of unsaturation and carbon number in the fatty acid chain. According to the analysis, Catfish oil consists of saturated fatty acid, monounsaturated fatty acid, and polyunsaturated fatty acids.

There are two types of omega-3 polyunsaturated fatty acid can be found in Catfish oil which are linoleic acid (ALA) and Eicosapentaenoic acid (EPA). Moreover, omega-6 polyunsaturated fatty acid also found in the Catfish oil. For example, Dihomo-gamma-linoleic acid, Eicosadienoic acid, and Gamma-linoleic acid. Then, the highest percentage of fatty acid composition found in Catfish oil are monounsaturated fatty acids which is 48.4136% before concentrated by lipase and 47.9947% after concentrate by lipase.Besides that, the lowest fatty acid composition percentage are polyunsaturated fatty acid which is 20.3218% before concentrate by lipase and 19.3249% after concentrate by lipase.

The result determine that after the Catfish oil concentrate by lipase, the amount of polyunsaturated fatty acid decrease. This might due to auto-oxidation and photo-oxidation of the oil. Research also prove that polyunsaturated fatty acids more likely to undergoes oxidation. When polyunsaturated undergoes oxidation will release hydroperoxide and this hydroperoxide further oxidizes will become ketones or aldehydes compound [13]. FTIR analysis in Table 1 shows the present of ketones functional group. Hence, this verified that the polyunsaturated fatty acid in the Catfish oil after concentrate by lipase reduce because of oxidation.

3.8 Acid value

The definition of acid value is the amount of potassium hydroxide in milligram need to neutralize the free fatty acid per one grams of oil. Thus, it can be used to predict the amount of free fatty acids present in the Catfish oil. Based on research, the lower the acid value, the higher the quality of oil. Food and Science department recommend the suitable acid value range of oil that can consumed by human was lower than 5mg KOH/g. There are many factors can influenced the acid value of oil such as extraction procedure, freshness of raw materials, and oil composition. The extraction procedure that used to extract the Catfish oil was optimum condition of enzymatic extraction. According to the data obtained, the acid value of Catfish oil extract by optimum condition of enzymatic extraction was 2.244mg KOH /g. This justify that the optimum condition of enzymatic extraction that optimize in the previous step was able to generate good quality of Catfish oil. After that the Catfish oil were undergoes lipase hydrolysis to separate and enrich the content of EPA and DHA in the Catfish oil. The acid value that we obtained from the experiment was 154.836mg KOH/g. This valid that lipase hydrolysis the bond between glycerol and fatty acids in triglyceride. The higher the acid value, the higher the amount of free fatty acid hydrolyses by lipase.

3.9 Peroxide value

Table 2. Peroxide value of catfish oil before and after concentrate by lipase.

Sample	Peroxide Value (Meq/g)
Catfish oil before concentrate by lipase	6.67
Catfish oil after concentrate by lipase	20.00

The definition of peroxide value is the amount of hydroperoxides in an oxidized oil. When oil undergoes oxidation will produce hydroperoxides. Based on the research, the lower the peroxide value, the better the quality of oil. There are many factors can cause the increased of oxidation rate of oil such as moisture content, fatty acid composition, light intensity, oxygen availability and temperature. A good quality of oil will have a peroxide value between 2Meq/g to 10Meq/g. The peroxide value that obtained from the optimum condition enzymatic extraction Catfish oil was 6.67Meq/g. This valid that the enzymatic extraction process able to produce good quality of oil and the temperature used does not faster the oxidation rate of Catfish oil.

However, after the Catfish oil concentrated by lipase the peroxide value increased up to 20Meq/g. Although the oil is incubated with nitrogen gas to prevent enter of oxygen gas and prevent oxidation but the incubation time is long, 24 hours. Some of the nitrogen gas might runoff and oxygen enter to triggers oxidation reaction. In addition, the more number of polyunsaturated fatty acid will cause fish oil extremely liable to oxidation. Hence, after lipase concentrate the Catfish oil increased the amount of polyunsaturated fatty acid, the peroxide value become higher.

4 Summary

Based on the screening data, the design verify that temperature, enzyme concentration, and incubation time have significant effect on the percentage of oil yields obtained by using enzymatic extraction process. The optimization result of this three parameter in the enzymatic extraction process were 0.91% enzyme concentration,51°C, and 155minute. The predicted solution shows that oil yields achieved in this condition is the maximum which is 16.8113±0.906%. Based on the spectra obtained in the FTIR analysis, Catfish oil before and after concentrate by lipase have many similar functional group such as carboxylic acid, alkenes, alkanes, ester, methyl, alcohol, phosphorous, and methylene group. However, the peaks of the ester bond decrease in the spectra after concentrate by lipase. This is because lipase hydrolyses the ester bond of triglycerides compounds.

In addition, that is the present of aldehydes group in the Catfish oil after concentrated by lipase. This verify that oxidation happens in Catfish oil. The quality of Catfish oil indicated by acid value test and peroxide value test. The acid value and the peroxide value of the Catfish oil before concentrated by lipase is 2.244mg KOH/g and 6.67Meq/g respectively. The acid value and the peroxide value are in the recommended range that justify the Catfish oil extract by using optimum condition enzymatic extraction is in a good quality. On the others hand, the Catfish oil after concentrated by lipase increased in acid value and peroxide value. Gas Chromatography-Mass Spectroscopy was used to analysis the fatty acid composition of Catfish oil. The analysis shows that there are two types of omega-3 polyunsaturated fatty acid found in Catfish oil which is EPA and ALA.

References

1. G. Elson, *"Feeding Catfish," in Catfish: Everything about Natural History*, Purchase, Health Care, Breeding, p. 95, (2003).
2. E. Teh, *"Fisheries in Malaysia: Can resources match demand?*," no. June, (2012).
3. *"Lutein + zeaxanthin and omega-3 fatty acids for age-related macular degeneration: the Age-Related Eye Disease Study 2 (AREDS2) randomized clinical trial.,"* JAMA, vol. 309, no. 19, pp. 2005–2015, May (2013).
4. R. J. W. Harry B. Rice, *"EPA and DHA Omega-3s for heart and brain health : The evidence speaks for itself*", **27**, no. June, pp. 38–41, (2016).
5. M. Russell L. Blaylock, *"DHA Supports Brain Development and Protects Neurological Function," Life Extension Magazine*, (2008). [Online]. Available: http://www.lifeextension.com/magazine/2008/1/report_dhafishoil/Page-01. [Accessed: 15-Mar-2017].
6. M. L. Juber, B. A., Jackson, K. H., Johnson, K. B., Harris, W. S., Baack, *"Breast milk DHA levels may increase after informing women: a community-based cohort study from South Dakota USA,"* Int. Breastfeed. J., (2016).
7. J. Eed,*"Factors affecting Enzyme Activity," Essai*, **10**, no. 10, pp. 48–51, (2012).
8. A. Bura Mohanarangan, *"Extraction of Omega-3 Fatty Acids From Atlantic Herring (Clupea harengus),"* no. November, (2012).

 https://doi.org/10.1051/mateconf/201818701005

9. C. Coury and A. M. Dillner, *"ATR-FTIR characterization of organic functional groups and inorganic ions in ambient aerosols at a rural site,"* Atmos. Environ., **43**, no. 4, pp. 940–948, (2009).
10. P. Arul Franco, N. Shenbagavinayagamoorthi, and A. Venkatesan, *"FT-IR determination of free fattyacids in Sardinella longiceps fish oil and its performance and emission characteristics in DI diesel engine,"* Int. J. Chem Tech Res., **6**, no. 7, pp. 3776–3783, (2014).
11. R. D. O`Brian, *"Fats Oils,"* p. 766.
12. J. Vongsvivut, M. R. Miller, D. McNaughton, P. Heraud, and C. J. Barrow, *"Rapid Discrimination and Determination of Polyunsaturated Fatty Acid Composition in Marine Oils by FTIR Spectroscopy and Multivariate Data Analysis,"* Food Bioprocess Technol., **7**, no. 8, pp. 2410–2422, (2014).
13. T. O. Akanbi, J. L. Adcock, and C. J. Barrow, "*Selective concentration of EPA and DHA using Thermomyces lanuginosus lipase is due to fatty acid selectivity and not regioselectivity,"* Food Chem., **138**, no. 1, pp. 615–620, (2013).
14. T. Ling, *"Oxidation of Polyunsaturated Fatty Acids and its Impact on Food Quality and Human Health,"* Adv. food Technol. Nutr. Sci., **1**, no. 6, p. 135–142., (2015).

Effect of Polymer Degradation on Polymer Flooding in Homogeneous Reservoirs

Xiankang Xin[1], Gaoming Yu[2], Ruicheng Ma[1], Keliu Wu[3] and Zhangxin Chen[1,3]

[1]*College of Petroleum Engineering, China University of Petroleum, Beijing, 102249 Beijing, China*
[2]*College of Petroleum Engineering, Yangtze University, 430100 Wuhan, China*
[3]*Department of Chemical and Petroleum Engineering, University of Calgary, T2N 1N4 Calgary, Canada*

Abstract. In this paper, physical and numerical simulations were applied to investigate the polymer degradation performance and its effect on polymer enhanced oil recovery (EOR) efficiency in homogeneous reservoirs. Physical experiments were conducted to determine basic physicochemical properties of the polymer, including viscosity, rheology, and degradation. A new mathematical model was proposed, and an in-house simulator was designed to further explore polymer degradation. The results of the physical experiments illustrated that polymer could increase polymer solution viscosity significantly, and the relationship between polymer solution viscosity and polymer concentration exhibited a clear power law relationship. However, the viscosity of a polymer solution with the same polymer concentration decreased with an increase in the shear rate, showing shear thinning performance. Moreover, the viscosity decreased with an increase in time, which was caused by polymer degradation. The validation of the designed simulator was improved when compared to the simulation results using ECLIPSE V2013.1 software. The difference between 0 and 0.1 day^{-1} in the polymer degradation rate showed a decrease of 6% in oil recovery after 2,000 days, according to simulation results, which demonstrated that polymer degradation had an adverse effect on polymer flooding efficiency.

1 Introduction

Polymers are closely related to our daily life. They have been widely applied in medicine, aerospace, the chemical industry and other fields [1–3]. Polymers have played a significant role, especially in the chemical industry. Polymer flooding, a mature chemical enhanced oil recovery (EOR) technology, has been used in oil field projects since the 1950s, where the viscosity of water was increased by adding polymer, and the water-oil mobility ratio was reduced [4]. The successful application of polymer flooding in oil and gas fields can alleviate the pressure from crude oil demand to a certain extent [5]. Therefore, polymer flooding is attracting increasing attention.

To make the best efficiency of polymer flooding, many researchers have done detailed research on the physicochemical properties of polymer; their studies mainly focused on the following aspects:
-Viscosity increasing. Many factors affect the viscosity of a polymer solution, such as temperature, polymer concentration, and the salinity of the water used in the solution preparation [6, 7].
-Rheological properties. A polymer solution often exhibits shear thinning within a certain range of shear rates [8, 9].
-Inaccessible pore volumes. An inaccessible pore volume is mainly influenced by permeability [10] and it affects the polymer flow in porous media [11].
-Absorption. Polymer adsorption depends on temperature, concentration, etc. [12].
-Degradation. Temperature, oxygen content and other factors affect polymer degradation [13, 14].

Polymer flooding simulation has also been carried out. With the continuous improvement of simulation research, most of the physicochemical properties of the polymer can be described in the simulation, but how to describe polymer degradation in a simulation has not been entirely solved. Some methods were used to solve this problem, but they did not work well. For example, because the application of the first order degradation model is based on concentration, it could not keep polymer concentration during polymer degradation [15].

In this paper, an experimental study was conducted to investigate some fundamental physicochemical properties of polymer including increasing viscosity, rheology, and degradation. A new mathematical model was proposed, and a novel in-house three-dimensional (3D) two-phase polymer flooding simulator was designed, with polymer degradation studies planned as part of the work scope. The validation of this simulator was done by comparing with the simulation results run in the ECLIPSE V2013.1 software. The simulations considering polymer degradation were undertaken using the designed simulator to investigate the influence of the polymer degradation on production indicators including a pressure difference and oil production. These results and simulations offer a tool to study the effects of polymer degradation on polymer flooding in homogeneous reservoirs.

2 Methodology

2.1 Physical experiments

2.1.1 Rheology test

The properties of polymer and the ion components and their concentrations in the brine are shown in Table 1 and Table 2. After preparation of the polymer solution including pre-shearing, the rheological experiment was conducted on the Physica MCR301 advanced rotary rheometer from Anton Paar. The viscosity of the polymer solution with different concentrations was tested at the shear rate of 6.7 s^{-1}, and the viscosity of the polymer solution with a concentration of 1,500 mg/L was measured at different shear rates at 45 °C.

Table 1. The properties of the polymer.

Properties	Description/Value
Type	Partially hydrolyzed polyacrylamide (HPAM)
Molecular weight	2.2×10^7
Solid content, wt.%	90
Hydrolysis degree, %	25
Filtration factor	1.6
Dissolution rate, hour	<2
Insoluble matter, wt.%	0.2
Granularity ≥1.0 mm, %	5
Granularity ≤0.2 mm, %	2.5

Table 2. Ion components and their concentrations in the brine.

Ion components	Concentration, mg/L
Na^+ and K^+	173.38
Ca^{2+}	46.92
Mg^{2+}	29.97
HCO_3^-	526.35
SO_4^{2-}	5.92
Cl^-	127.46
TDS	910.00

2.1.2 Polymer degradation experiment

After being deoxidized, the prepared polymer solution with a concentration of 1,500 mg/L was sealed and placed in a thermotank at a temperature of 45 °C. The viscosity of polymer solutions was measured at 45 °C at 1 day, 5 day, 15 day, 20 day, 40 day, 60 day, 80 day,100 day and 120 day.

2.2 Mathematical model

The underlying assumptions are as follows.
- The flow was isothermal and followed Darcy's law.
- The model was 3D with two phases (oil and water).
- The rock was compressible and anisotropic, and the fluid was compressible.
- Polymer components only included high and low molecular weight polymer components. The viscosity of the polymer solution was determined by the high molecular weight polymer component.
- The virgin polymer solution polymer contained only the high molecular weight polymer component.
- The mixture of water and polymer is ideal, and they existed only in the water phase.
- The effects of the capillary force and gravity were taken into account.

According to the conservation of mass, the mass conservation equations of all components were obtained as follows.

For oil:

$$\nabla\cdot\left[\frac{\vec{k}k_{ro}}{\mu_o B_o}\left(\nabla p_o-\rho_o g\nabla D\right)\right]+q_o=\frac{\partial}{\partial t}\left(\frac{\phi s_o}{B_o}\right) \tag{1}$$

For water:

$$\nabla\cdot\left[\frac{\vec{k}k_{rw}}{\mu_{we}B_w R_k}\left(\nabla p_w-\rho_w g\nabla D\right)\right]+q_w=\frac{\partial}{\partial t}\left(\frac{\phi s_w}{B_w}\right) \tag{2}$$

For high molecular weight polymer component:

$$\begin{aligned}&\nabla\cdot\left[\frac{\vec{k}k_{rw}c_{hp}}{\mu_{we}B_w R_k}\left(\nabla p_w-\rho_w g\nabla D\right)\right]-c_{hp}R_{pd}+q_w c_{hp}\\&=\frac{\partial}{\partial t}\left(\frac{\phi(1-f_{ip})s_w c_{hp}}{B_w}\right)+\frac{\partial[(1-f_{ip})(1-\phi)\rho_r c_{hpa}]}{\partial t}\end{aligned} \tag{3}$$

For low molecular weight polymer component:

$$\begin{aligned}&\nabla\cdot\left[\frac{\vec{k}k_{rw}c_{lp}}{\mu_{we}B_w R_k}\left(\nabla p_w-\rho_w g\nabla D\right)\right]+c_{hp}R_{pd}+q_w c_{lp}\\&=\frac{\partial}{\partial t}\left(\frac{\phi(1-f_{ip})s_w c_{lp}}{B_w}\right)+\frac{\partial[(1-f_{ip})(1-\phi)\rho_r c_{lpa}]}{\partial t}\end{aligned} \tag{4}$$

in which, ∇ is the Hamilton operator, which represents the gradient of the subsequent quantity; $\vec{k}$ is the absolute permeability tensor in um^2; k_{ro} and k_{rw} are the relative permeabilities of oil and water; μ_o is the viscosity of the oil phase in mPa·s; μ_{we} is the effective viscosity of the water phase in mPa·s; B_o and B_w are the oil and water formation volume factor in m^3/m^3; p_o and p_w are the pressures of the oil and water phase in Pa; ρ_o and ρ_w are the oil and water densities in kg/m^3, and g is the gravitational acceleration in m/s^2. D is the vertical height in m; q_o and q_w are the source/sink terms for oil and water phase in $m^3/(day\cdot m^3)$; ∂ is the symbol used to denote partial derivatives, and t is the time in day. ϕ is the porosity; s_o and s_w are the oil and water phase saturation. The permeability reduction factor $R_k = 1 + (RRF - 1) \times C_{pa}/C_{pa\max}$, where RRF is the residual resistance factor; c_{pa} and $c_{pa\max}$ are the polymer adsorption concentration and maximum polymer adsorption concentration in kg/m^3. c_{hp} and c_{lp} are the high, and low molecular weight polymer concentrations in kg/m^3; R_{pd} is the polymer degradation rate in day^{-1}, and f_{ip} is the inaccessible pore volume factor. ρ_r is the rock density in kg/m^3, while c_{hpa} and c_{lpa} are the adsorption concentrations of the high and low molecular weight polymer in kg/m^3.

In addition to the mass conservation equations, the auxiliary equations and state equations are as follows. The auxiliary equations include:

$$s_o + s_w = 1 \tag{5}$$

$$p_{cow}(s_w) = p_o - p_w \tag{6}$$

where $p_{cow}(s_w)$ is the capillary pressure in the water-oil system in Pa, which is a function of water phase saturation. The state equations include:

$$k_{ro} = k_{ro}(s_w) \tag{7}$$

$$k_{rw} = k_{rw}(s_w) \tag{8}$$

$$\rho_o = \rho_o(p_o) \tag{9}$$

$$\rho_w = \rho_w(p_w) \tag{10}$$

$$\phi = \phi(p_r) \tag{11}$$

$$\mu_{wc} = \mu_{wc}(c_{hp}, v_p) \tag{12}$$

where p_r is the reservoir pressure in Pa, V_p is the velocity of polymer solution in m/day. Equation (12) shows that the effective viscosity of the water phase is not the sole function of the high molecular weight polymer component but is also a function of the polymer solution velocity. Therefore, the relationship between the viscosity of polymer solution and shear rate, one of the rheological properties of polymer solution, can be described in the mathematical model.

Meanwhile, it can be seen from the assumptions, mass conservation equations and Equation (12) that the high molecular weight polymer component was only degraded, and the total polymer concentration retained conservation and followed the mass conservation. This model conformed to the theory of polymer chemistry.

To ensure the stability of computation, the full implicit method was applied to solve the mass conservation equations under the boundary and initial conditions. More details about the fully implicit method can be found in [16]. Finally, pressure was obtained along with production, oil and water phase saturation distribution, as well as the high and low molecular weight polymer concentration.

3 Results and discussion

3.1 Rheology of polymer

The relationship between the viscosity of the polymer solution and polymer concentration is indicated in Figure 1 where their relationship curve shows an excellent power law relationship, which can be written as:

$$\mu_{ps} = 0.0001 \times c_p^{1.7782} \tag{13}$$

where μ_{ps} is the viscosity of the polymer solution in mPa·s, and c_p is the polymer concentration in mg/L. The viscosity of the polymer solution increased with an increase in polymer concentration. The main reason is that the formation of a larger hydrodynamic radius after polymer hydration increases the viscosity of polymer solution with a higher polymer concentration [17]. However, the viscosity of the polymer solution with a concentration of 1,500 mg/L decreases with an increase in shear rate, as shown in Figure 2. The main reason for shear thinning of the polymer solution is that the higher shear rate makes the intertwined polymer molecules collapse, which means that the viscosity of the polymer solution will decrease [18].

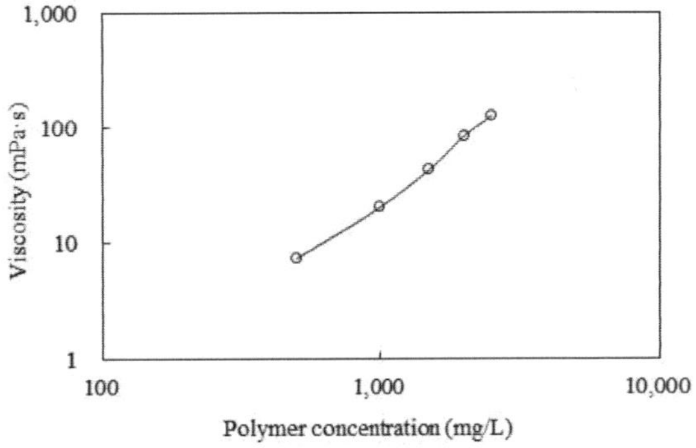

Figure 1. The relationship between the viscosity of the polymer solution and polymer concentration.

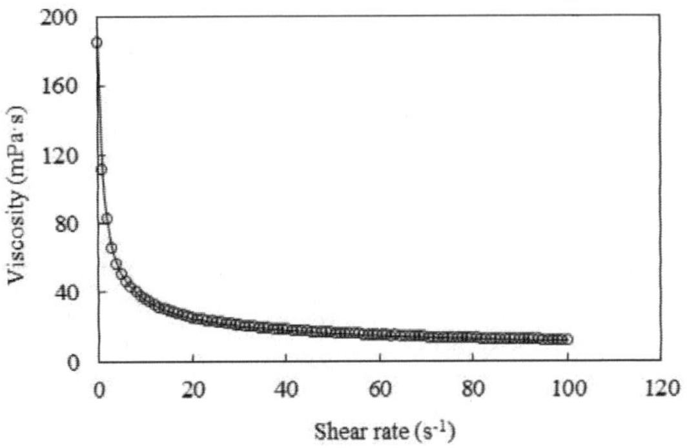

Figure 2. Shear thinning of polymer solution with a concentration of 1,500 mg/L.

3.2 Polymer degradation

Polymer degradation is illustrated in Figure 3. The viscosity of polymer solution decreases with an increase in time. The process of polymer degradation breaks down the long molecular polymer chains; hence, the reduction in length of average molecular chains leads to a decrease in the viscosity of polymer solution [19, 20]. Unlike the relationship between the viscosity of polymer solution and polymer concentration, the polymer degradation curve shows an excellent exponential relationship, which can be written as:

$$\mu_{ps} = 38.156 \times e^{-0.002t_{pd}} \tag{14}$$

where t_{pd} is the polymer degradation time in day. Combining the polymer rheological experiment and polymer degradation experiment results, the computational expression of the polymer degradation rate is given:

$$R_{pd} = -\frac{dc_{hp}/c_{hp}}{dt_{pd}} \tag{15}$$

where d is the derivative symbol. Here, the polymer degradation rate can be calculated when its concentration is 1,500 mg/L, which is about 0.001 day^{-1}.

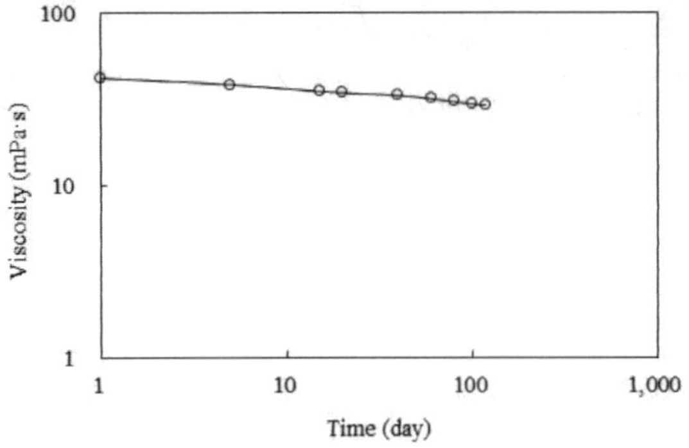

Figure 3. Polymer degradation curve.

3.3 Simulation

3.1.1 Validation

To validate the accuracy of our designed simulator, the simulation results of ECLIPSE V2013.1 software were used to compare with that of our designed simulator because it is a recognized numerical reservoir simulator, and its simulation result is authoritative. Their main input parameters, relative permeabilities, and capillary pressures were the same, which are provided in Table 3 and Table 4. Figure 4 indicates the comparison result of production indicators including pressure differences, oil production, water production, water cut, cumulative oil production, cumulative water production and oil recovery. By comparison, it is clear that the simulation results were very close, the difference between each production indicator was less than 0.2%. The remaining oil saturation distribution after 1,500 days of monitoring by ECLIPSE V2013.1 software and our designed simulator are presented in Figure 5. From the data on the overall remaining oil distribution, the results of both the ECLIPSE V2013.1 software and our designed simulator were very similar. Thus, the accuracy and high reliability of our designed simulator was confirmed.

Table 3. Main input parameters for simulation.

Parameters	Value
Number of blocks along x, m	21
Number of blocks along y, m	21
Number of blocks along z, m	3
Length of the block along x, m	1.4236
Length of the block along y, m	1.4243
Length of the block along z, m	1.3967
Initial porosity, fraction	0.2423
Initial permeability in x-direction, mD	400
Initial permeability in y-direction, mD	400
Initial permeability in z-direction, mD	40
Rock compressibility, MPa^{-1}	2.7×10^{-3}
Rock density, Kg/m^3	2,600
Stock tank oil density, Kg/m^3	880
Initial oil viscosity, mPa·s	8.90
Oil compressibility, MPa^{-1}	1.2×10^{-3}
Oil formation volume factor	1.068
Initial water density, Kg/m^3	1,000
Water viscosity, mPa·s	0.69
Water compressibility, MPa^{-1}	4.26×10^{-4}
Water formation volume factor	1.016
Initial reservoir pressure, MPa	2
initial water saturation, fraction	0.3172
initial oil saturation, fraction	0.6828
Polymer concentration, mg/L	1,500
Polymer degradation rate, day^{-1}	0
Inaccessible pore volume factor, fraction	0.05
Maximum polymer absorption, $K_g/K_{g\ rock}$	1.60×10^{-5}
Residual resistance factor	1.30
Bottom hole pressure of production wells, MPa	2
Injection rate, m^3/day	0.65

Table 4. Relative permeabilities and capillary pressures for simulations.

Sw	Krw	Krow	Pcow
0.3061	0.0000	1.0000	75349
0.3372	0.0001	0.8762	72491
0.3563	0.0017	0.8014	67368
0.3834	0.0031	0.7020	64853
0.4145	0.0055	0.5864	61115
0.4416	0.0099	0.4884	59305
0.4727	0.0150	0.3837	56567
0.5013	0.0231	0.2966	53832
0.5289	0.0381	0.2231	52450
0.5580	0.0612	0.1551	50505
0.5881	0.0912	0.0952	48877
0.6162	0.1293	0.0599	47696
0.6473	0.1810	0.0299	45611
0.6744	0.2272	0.0122	44483
0.7052	0.2884	0.0000	43850

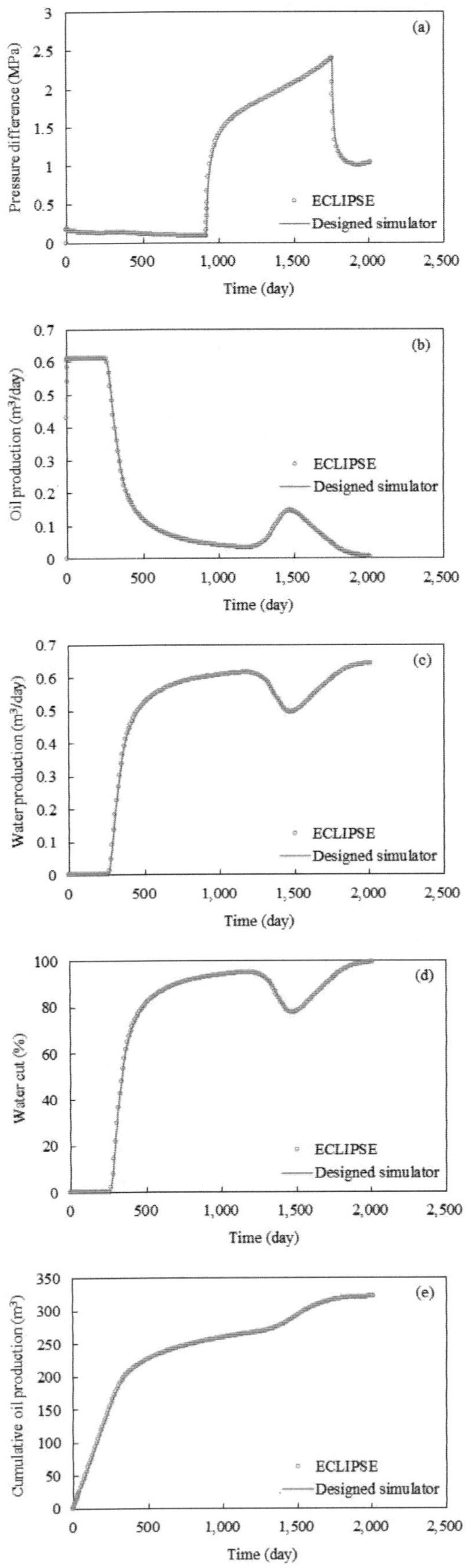

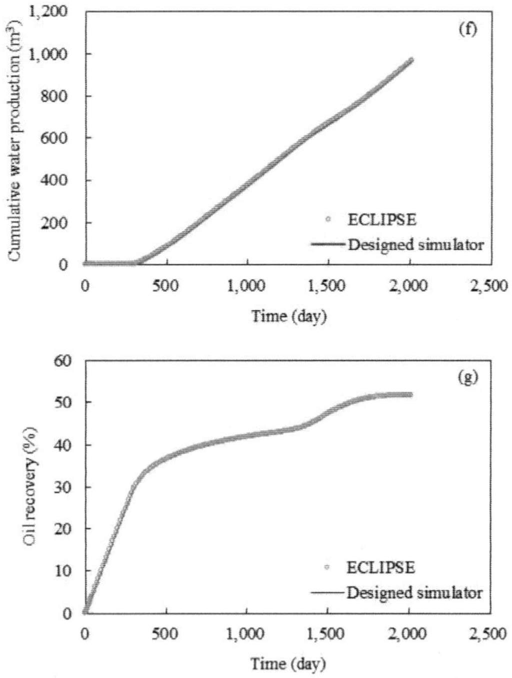

Figure 4. Comparison results of (a) pressure difference, (b) oil production, (c) water production, (d) water cut, (e) cumulative oil production, (f) cumulative water production, and (g) oil recovery of ECLIPSE V2013.1 software and designed simulator.

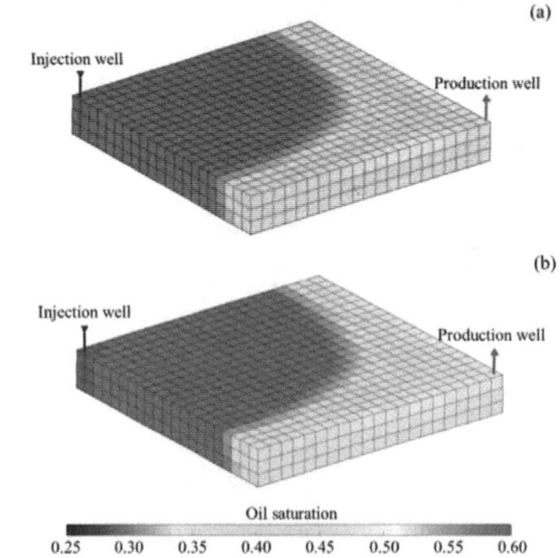

Figure 5. Comparison results of remaining oil saturation distribution after 1,500 days of (a) ECLIPSE V2013.1 software and (b) the designed simulator.

3.1.2 Effect of polymer degradation

Four more simulations were undertaken to analyze the effect of polymer degradation on production indicators. They considered the polymer degradation rate of 0.001 day^{-1}, 0.002 day^{-1}, 0.010 day^{-1} and 0.100 day^{-1}. The rest of the parameters were the same. Figure 6 provides the comparison result of production indicators at different polymer degradation rates. This figure demonstrates that the higher polymer degradation rate could cause a more

 https://doi.org/10.1051/matecconf/201818701006

significant impact on production indicators. The reduction of production indicators caused by polymer degradation after 1,500 days is shown in Table 5. Figure 7 shows the remaining oil saturation distribution after 1,500 days of simulation considering the polymer degradation rate of 0.100 day^{-1}. From this figure, the remaining oil saturation of the simulation considering the polymer degradation rate of 0.100 day^{-1} was more than that of the simulation considering the polymer degradation rate of 0 day^{-1}, as seen in the comparison between Figure 5 and Figure 7. Notably, the oil recovery, one of the most critical production indicators, decreased 6% after 2,000 days.

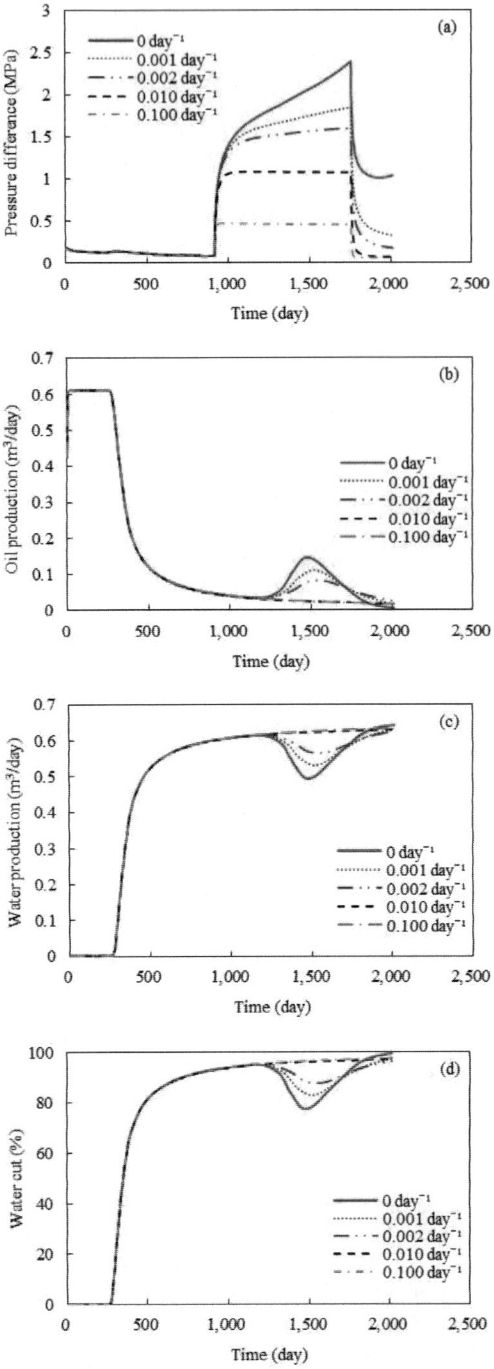

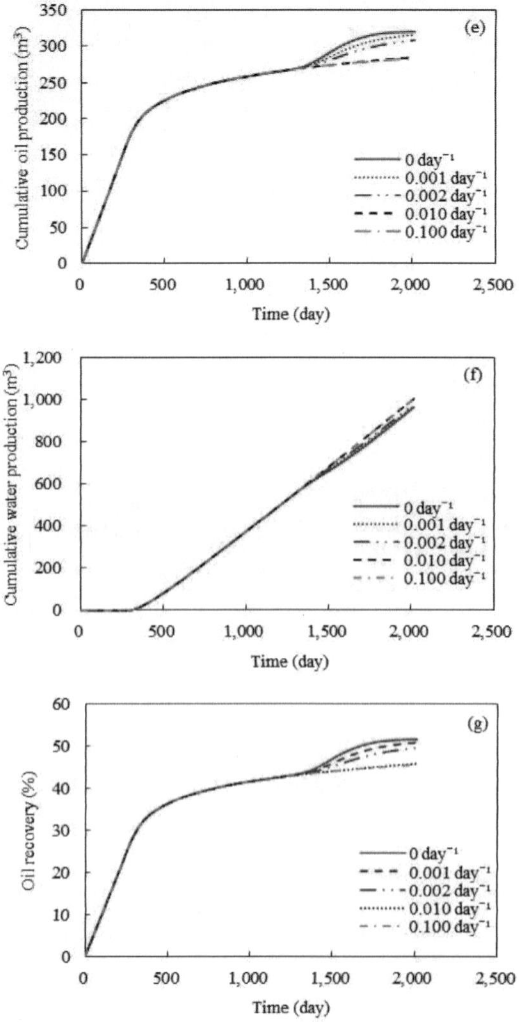

Figure 6. Comparison results of production indicators including (a) pressure difference, (b) oil production, (c) water production, (d) water cut, (e) cumulative oil production, (f) cumulative water production, and (g) oil recovery at different polymer degradation rates.

Table 5. Reduction of production indicators caused by polymer degradation after 1,500 days.

Production indicators	**Polymer degradation rate, day^{-1}**			
	0.001	**0.002**	**0.010**	**0.100**
Pressure difference, MPa	0.32	0.51	0.99	1.61
Oil production, m^3/day	0.03	0.06	0.12	0.12
Water production, m^3/day	-0.03	-0.07	-0.13	-0.13
Water cut, %	-5.07	-10.04	-18.46	-18.78
Cumulative water production, m^3	6.55	11.09	18.25	18.55
Cumulative oil production, m^3	-7.24	-12.22	-20.05	-20.41
Oil recovery, %	1.06	1.79	2.94	2.99

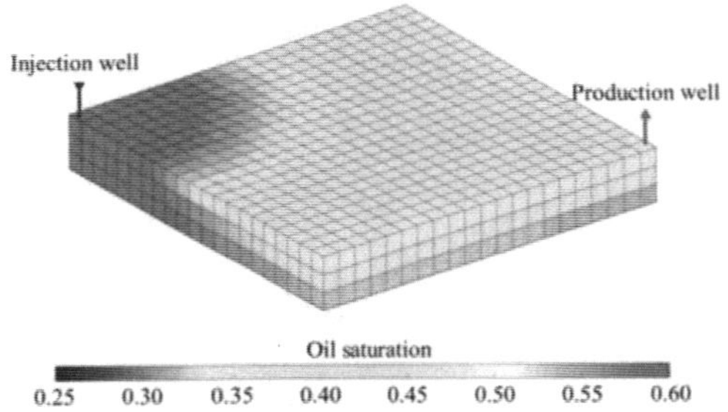

Figure 7. Remaining oil saturation distribution after 1,500 days of the simulation considering the polymer degradation rate of 0.100 day^{-1}.

4 Conclusions

The primary objective of this paper was to investigate the effect of polymer degradation on polymer flooding in a homogeneous reservoir to provide a theoretical guide for the development of similar oilfield reservoirs. The results of the physical experiment indicate that the relationship between the viscosity of the polymer solution and polymer concentration had a good power law relationship; the viscosity of the polymer solution increased with the increase of the polymer concentration. However, it decreased with an increase in shear rate; the rheological property of polymer solution indicated shear thinning performance. Moreover, the polymer degradation led to a viscosity reduction of polymer solution with the increase of time, and there was an excellent exponential relationship. Based on the physical experiment and mathematical model proposed in this paper, the polymer degradation rate was introduced and calculated. Furthermore, the 3D two-phase polymer flooding simulator that considered the polymer degradation was designed and validated with high accuracy and reliability when compared to the simulation results run in ECLIPSE V2013.1 software. The results of the simulation considering the polymer degradation indicates that the polymer degradation significantly influenced the production indicators. In the simulation considering the polymer degradation rate of 0.100 day^{-1}, the recovery factor after 2,000 days was 6% lower than that in the simulation without considering the polymer degradation. To minimize the impact of the polymer degradation on polymer flooding, some methods have been proposed [4, 21], which should be followed by a reduction in polymer degradation to increase polymer flooding efficiency.

This work was supported by the National Science and Technology Major Project of China (No.2016ZX05025-001-005)

References

1. Q. Wei, X. Zhang, H. Wang, Curr. Opin. Colloid. Interface. Sci., **35**, 36-41 (2018)
2. X. Zhang, Y. Chen, J. Hu, Pro. Aerosp. Sci., 97, 22-34 (2018)
3. R. Liu, W. Pu, D. Du, J. Gu, L. Sun, J Petro. Sci. Eng, **164**, 467-484 (2018)
4. D.C. Standnes, I. Skjevrak, J Petro. Sci. Eng, **122**, 761-775 (2014)
5. X. Xin, Y. Li, G. Yu, W. Wang, Z. Zhang, M. Zhang, W. Ke, D. Kong, K. Wu, Z. Chen, Energies, **10** (2017), doi:10.3390/en10111698
6. A.A. Alquraishi, F.D. Alsewailem, Carbohydr. Polym., **88**, 859-863 (2012)
7. M. Algharaib, A. Alajmi, R. Gharbi, J Petro. Sci. Eng, **115**, 17-23 (2014)
8. J. Choi, D. Ka, T. Chung, J. Jung, G. Koo, T. Uhm, S.H. Jung, S. Park, H.T. Jung, Macromolecular Research, **23**, 518-524 (2015)
9. S. Cai, X. He, K. Liu, R. Zhang, L. Chen, Iran. Polym. J, **24**, 663-670 (2015)
10. Y. Ma, J. Hou, Oilfield Chemistry, **34**, 361-365 (2017)
11. A. Li, H. Song, H. Xie, Petroleum Geology and Recovery Efficiency, **23**, 70-74 (2016)
12. G. Zhao, J. Fang, B. Gao, Y. Wang, A. Chen, D. Wang, C. Dai, Oilfield Chemistry, **32**, 62-66 (2015)
13. J. Wang, H. Liu, J. Ind. Eng. Chem, **20**, 656-667 (2014)
14. E. Unsal, A. Berge, D.A.Z. Wever, J Petro. Sci. Eng, **163**, 671-682(2017)
15. B.I. Choi, M.S. Jeong, K.S. Lee, Polym. Degrad. Stab, **110**, 225-231 (2014)
16. Z. Chen, G. Huan, Y. Ma, *Computational Methods for Multiphase Flows in Porous Media* (Society for Industrial and Applied Mathematics, Philadelphia, USA, 2006)
17. D. Fang, R. Guo, H. Hua, *Acrylamide polymer* (Chemical Industry Press, Beijing, 2006)
18. G. Lu, N. Shao, M. Cao, B. Wang, Engineering Plastics Application, **44**, 90-92 (2016)
19. A.M. Mansour, R.S Al-Maamari, A.S Al-Hashmi, A. Zaitoun, H. Al-Sharjic, J Petro. Sci. Eng, **115**, 57-65 (2014)
20. L.F.L. Oliveira, D.J. Schiozer, M. Delshad, J Petro. Sci. Eng, **147**, 346-355 (2016)
21. L.J. Giraldo, M.A. Giraldo, S. Llanos, G. Maya, R.D. Zabala, N.N. Nassar, C.A. Franco,V. Alvarado, F.B. Cortés, J Petro. Sci. Eng, **159**, 841-852 (2017)

MATEC Web of Conferences **187**, 02001 (2018) https://doi.org/10.1051/matecconf/201818702001
ICCMP 2018

Development of Rigid Polyurethane Foam (RPUF) for Imitation Wood Blown by Distilled Water and Cyclopentane (CP)

Ratchanon Boonachathong[1], Bordin Kaewnok[2], Halim Widjaja[2], and Suksun Amornraksa[1]

[1] *the Department of Mechanical and Process Engineering, The Sirindhorn International Thai-German Graduate School of Engineering, King Mongkut's University of Technology North Bangkok, Bangkok, Thailand*
[2]*Technical Service and Development Department, Covestro, PUR system house, Samut Prakan, Thailand*

Abstract. Rigid polyurethane foam (RPUF) used for imitation wood is typically prepared by using 1-dichloro-1-fluoroethane (HCFC-141b) as a blowing agent. However, this chemical is a hydrofluorocarbon which severely causes ozone depletion to the atmosphere. In this study, a more environmental-friendly RPUF was prepared by using distilled water and cyclopentane (CP) as alternative blowing agent. Several properties of the prepared RPUF were investigated and measured such as density (kg/m^3), surface hardness (Durometer, type D), and glass transition temperature (°C) using differential scanning calorimeter (DSC). It was found that when the amount of blowing agents decreased, the foam density was increased as well as the surface hardness. The developed RPUF with CP co-blown has higher surface hardness and glass transition temperature compare to pure water-blown RPUF at the same density (353 kg/m^3). And the new RPUF produced has a good potential to substitute for a conventional RPUF.

1 Introduction

Rigid polyurethane foam (RPUF) is a very useful thermoset polymer. It has been used in several applications, for example, automotive parts, appliance and construction materials, insulation, e.g. freezers, imitation woods and decoration products [1]. For imitation wood, there are several product properties that are of important concerns such as strength, thermal stability, and light weight.

RPUF is typically produced by reaction of polyol and Methane diphenyl diisocyanate (MDI). The reaction occurs simultaneously and yields polyurethane foam and carbon dioxide gas as products. The reaction is exothermic. The chemical structures of MDI, polyol and chemical reactions are illustrated in the Figure 1 and Figure 2 [2].

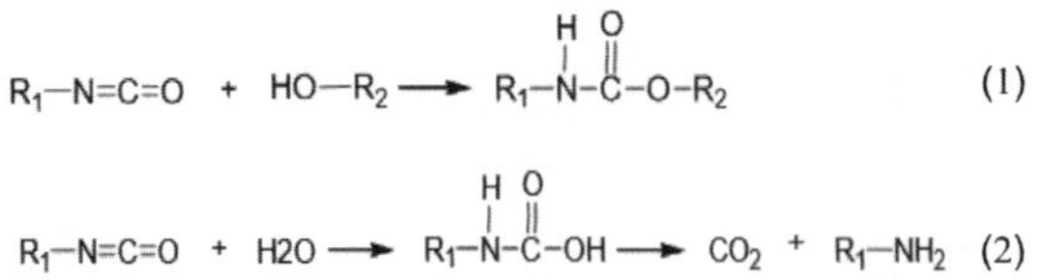

Figure 1. Primary reactions: polymerization (condensation) and foaming reaction of RPUF produced urethane and amine with carbon dioxide reactions (1) and (2) respectively.

Two main reactions for polyurethane formation are polymerization and foaming reaction. For polymerization, the isocyanate (MDI) reacts with polyol and forms urethane linkage (1). At the same time, MDI reacts with water and forms unstable carbamic acid which decomposed to carbon dioxide (CO_2) and amine in foaming reaction (2) [3].

Then, products of primary reactions can continue reacting with MDI.

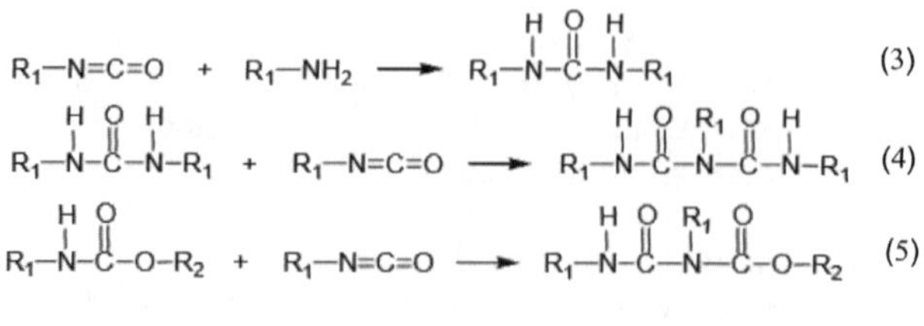

Figure 2. Secondary reactions of produced RPUF prouduced urea, biuret and allophanate; reactions (3), (4) and (5) respectively.

Amine reacts with MDI and produces urea. This urea forms hard segment phases, polyuria (3). Urea reacts with MDI and forms Biuret (4). During the rising of temperature from exothermic reaction, urethane from equation 1 may also react with excess quantity of MDI and produce allophanate (5). This allophanate is considered a crosslinker in RPUF [4].

The obtained rigid polyurethane foam has close cell structure [5]. The cellular structure of rigid foam is formed by means of chemical and physical blowing agent. An example of chemical blowing agents for RPUF is distilled water. It reacts with MDI and generates polyuria, polybiuret and carbon dioxide as described by (2)-(4). The produced gas bubbles create cellular structure of the rigid foam. Another method to produce foam structure is via the evaporation of physical blowing agents. The evaporation is promoted by exothermic heat from the reaction. There are also other blowing agents that can be used in RPUF production, depending on required product properties and costs.

Blowing agent is a substance that is able to produce a cellular structure via a foaming process in a variety of materials that undergo hardening or phase transition, such as polymers, plastics, and metals. Blowing agent occupy 95–97% of the foam volume. It contributes around 50-70% to the overall foam conductivity [6]. Hence, it has significant influence on insulation property.

Traditionally, auxiliary blowing agent that is widely used in production of polyurethane foam is trichlorofluorocarbon (CFC-11). However, according to the Montral protocol in 1987, chlorofluorocarbons (CFCs) blowing agent was banned in industrial due to contained Halogens which is harmful to the ozone layer, and consequently contributes to global warming and other environmental problems. So, its utilization is gradually decreased. In addition, the Kigali amendment referring to Montreal protocol states that 1-Dichloro-1-fluoroethane (HCF-141b) has to be replaced (phase-out) due to the depletion of ozone layer. The main context of the amendment is to gradually reduce 80-85% consumption of HFC substances used by 2040. Moreover, the Copenhagen Amendment outlines the phase-out allows HCFC consumption between 1996 and 2040 [7], [8].

The search for alternative blowing agent to replace CFCs in polyurethane foams has been a challenge due to numbers of important factors. Concerned factors are environmental consideration; air pollution, feasibility consideration; toxicity, flammability, compatibility with materials of construction, safety and economic manufacturing process, etc. Several studies showed that the selection of blowing agent is very crucial as it can directly affect physical and thermal properties of RPUF. So, many alternative blowing agents for polyurethane foam have been studied base on available formulation and production technology to overcome those challenges.

There are numbers of studies that are associated with water-blown foam, e.g., foam stability, flow ability, foam strength and etc. Water (H_2O) is a non-ozone depleting, non-toxic and affordable blowing agent. Researches on water blown rigid foams and their applications have performed investigation on water level and its effect to rigid foam. The studies found that water is an important part in determining the RPUF density (kg/m^3). The density of RPUF can be increased by decreasing the amount of water [9, 10].

Cyclopentane (C_5H_{10}) is an alicyclic hydrocarbon. It has a boiling point at 49.2 °C. It behaves as physical blowing agent in RPUF. When temperature is rising due to heat of reaction, CP will evaporate and create foam structure. In a study by Li Xiaobin et al., they evaluated and compared the performance of water and cyclopentane. Their results showed that the reaction times of CP full blown are longer compared to waterborne recipe [11].

The present of rigid structures i.e. aromatic rings of MDI and increasing of crosslink density in RPUF lead to increase of glass transition temperature (Tg). Studies showed that higher glass transition temperature is possible when increasing amount of MDI and water in during product formulation [12-14]. Moreover, when distilled water is used the mobility of polymer chain can be restricted and results in increasing of glass transition temperature [15-17].

In this study, the composition of isocyanate (MDI) used in the preparation of RPUF is the same quantity (same ratio of ROH: RNCO). The performance of RPUF prepared from distilled water and cyclopentane blowing agents are investigated. Cyclopentane will be used as co-blowing agent (CP co-blown) with distilled water. Key properties for the imitation wood RPUF product are measured. They are density (kg/m^3), surface hardness (shore D), and glass transition temperature (Tg). Materials and characterization methods used for the RPUF prepared from distilled water and cyclopentane as blowing agent are described in the following sections.

2 Experiments

2.1 materials

The chemicals used for preparation of RPUF samples in this study were supplied by Covestro, polyurethane system house, Thailand (former Bayer material science). Methane diphenyl diisocyanate (MDI) used has an average functionality of 2.7 and characterized by NCO%= 30.3 – 31.2, viscosity at 25 C is 210 – 230 mPa s.

Polyol-1 is a long chain polymer. It is propylene glycol initiator with average functionality 2, hydroxyl value 113-120 mgKOH g^{-1}. The viscosity at 25 °C is 150 mPa s.

Polyol-2 is a short chain polymer. It is sucrose initiator with average functionality 3-4, hydroxyl value 350-360 mgKOH g-1. The viscosity at 25 °C is 3000 mPa s.

Catalyst-1 contains 33% of triethylene Diamine (TEDA) and 67% of dipropylene glycol. It is used to promote gel reaction.

Catalyst-2 is methyl imidazolium. It is used to improve adhesion.

Silicone is used as foam stabilizer. It emulsifies the mixing of RPUF and creates fine cell foam size.

Glycerin ($C_3H_8O_3$) is a plain polyol compound containing 3 hydroxyl groups. It has a short chain structure which can increase level of crosslink in RPUF. Blowing agent-1, 2 are distilled water (generated in our laboratory) and cyclopentane (CP).

2.2 Preparation of rigid polyurethane foam samples

The laboratory works includes formulation, free rise hand foam and mold tests (20 cm x10 cm x 2 cm). The RPUF were prepared by hand mixing technique, by mixing polyol with catalysts 1, 2, and additives together. Once well-mixed, blowing agents 1 and 2 were added into the formulated polyol (circa. 20-25 °C) and mixed for 6 seconds. The stirrer speed was set at 3000 rounds per minute (rpm) for entire experiment.

The reaction times (second) for creaming, gelling and tracking free times were observed via free rise hand foam in plastic cup (200 ml). The amounts of blowing agents, distill water and cyclopentane is varied to obtain the desired reaction times and density (kg/m^3).

Then, properties of rigid foam sample will be examined via mold test. Mold temperature before pouring the mixed chemicals is controlled at 40-42 °C. During mold test, density of the sample was varied by different input weights of mixed chemicals. The mixed chemicals were poured into the constant mold volume. The mold was closed for 10 minutes before being opened to measure density.

The composition of raw materials with different blowing agents used in this work to prepare RPUF are summarized and shown in Table 1.

Table 1. Chemical compositions of RPUF, prepared at 20°C.

Composition (pphp)	Recipe			
	1	**2**	**3**	**4**
Polyol 1 and 2	100	100	100	100
Glycerin	1.5	1.5	1.5	1.5
Silicone	1	1	1	1
Catalyst 1 and 2	1.2	1.2	1.2	1.2
Distilled water	0.63	0.5		
Cyclopentane		0.5	2.46	1.23

Isocyanate index 105
pphp: parts per 100 parts of polyol by weight (g)
Mixing ratio; ROH: RNCO = 1:1

The quantity of blowing agent used in this study is controlled base on mole of produced blowing gas. From reaction (1), considering stoichiometry, one mole of water reacted with MDI will produce one mole of carbon dioxide. Considering the recipe 2 (CP co-blown), 0.5 gram of water produced circa. 0.028 mole of gas (CO2). Likewise, 0.5 gram of CP produced 0.00713 mole of gas (CP). The total produced gas is 0.03513 mole. Thus, for the recipe 1 (full waterborne), the used quantity of water is circa. 0.6324 gram. So, the same amount of gas (mole) can be produced. In case of recipe 3 (full CP blown), it required circa. 2.46 g accordingly. Practically, high amount of CP is not preferred due to flammability concerns. So, in this study, we did not do thermal analysis for recipe 3 and 4. However, the comparisons of reaction times are still observed and shown in next section.

Prepared 10 RPUF samples of recipe 1 and 2 have density between 218 – 353 kg/m^3. The density is controlled by weight. The RPUF samples are left for 24 hours to ensure that the cure is stable before measuring surface hardness (shore D). The average value of the surface hardness from upper and lower side of the RPUF is then measured.

Finally, the sample recipe 1 and 2 with density of 353 kg/m^3 were analyzed for thermal property by using DSC.

2.3 Analytical methods

Density (kg/m^3) was measured according to ASTM D 1622-8 standard. The RPUF sample densities were calculated via mass over volume. The RPUF samples were prepared by mold (size 10 cm x 20 cm x 1 cm). In this study, prepared samples have density between 218 to 353 kg/m^3.

Surface hardness is measured using Durometer tester, Teclock GS-707 (Shore D), DIN ISO 7619-1 standard. The RPUF samples were prepared by mold (10 cm x 20 cm x 1 cm). After 12 hours, the surface hardness of the sample is measured.

Differential scanning calorimeter (DSC Q2000 V24.11) was used to measure glass transition temperature (T_g) of rigid foam samples. The samples were investigated at nitrogen flow rate of 60 mL/min, heating rate of 15 °C per minute to 300 °C.

3 Results and discussions

3.1 Reaction times of RPUF using blowing agents, distilled water and cyclopentane

From prepared chemicals compositions of study RPUF, the reaction times are shown in Table 2.

Table 2. Reaction times of RPUF using blowing agent, distilled water and cyclopenthane.

Composition (pphp)	1	2	3	4
Cream time (s)	41	37	39	65
Gel time (s)	89	80	96	96
Tack free time (s)	119	105	165	121
Free rise density (kg/m^3)	166	136	52.7	222

Isocyanate index 105
pphp: parts per 100 parts of polyol by weight (g)
Mixing ratio; ROH: RNCO = 1:1

RPUF prepared by recipe 1 (pure water blown) had a shorter reaction times compared to cyclopentane (CP) full blown in recipe 3 and 4 due to the effect of exothermic heat reaction. The produced heat causes the reaction move forward faster. So, it takes shorter time to complete the reaction. Meanwhile, recipe 3 and 4 required longer times to complete the reaction. The reason is that CP in those recipes requires longer time to absorb heat in evaporation. Recipe 3 has longer reaction times compared to recipe 4 due to higher amount of CP. The presence of higher amount of CP results in longer reaction times regarding the absorption heat, to change from liquid to vapor.

It can be seen that RPUF prepared by recipe 2, CP co-blown yielded proper reaction times. From the viewpoint of commercial scale of RPUF production, the reaction times of this recipe are suitable for up-scaling.

3.2 Effect of blowing agents on density (kg/m^3) and surface hardness (shore D) of rigid foams

In this work, the averages of upper and lower side of the RPUF surface were measured. The relationship of density and surface hardness using distilled water as a chemical blowing agent (recipe 1) is illustrated in Figure 3 below.

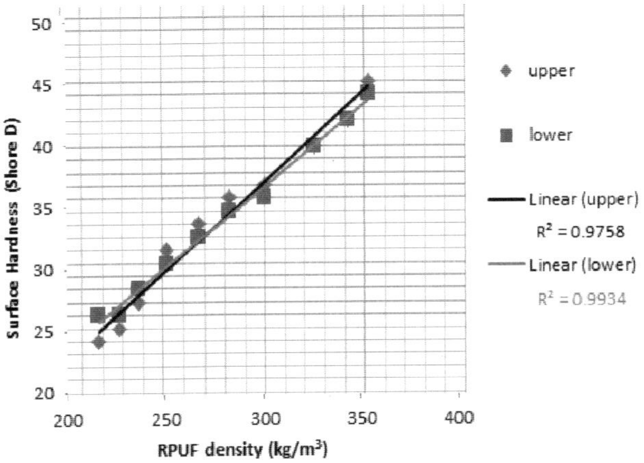

Figure 3. Relationship of density (kg/m^3) and surface hardness (shore D) using distilled water as a blowing agent.

Considering the increased density from 218 to 353 kg/m3, the surface hardness (shore D) is increased both on the upper and lower side of the sample. The average surface hardness on the upper and lower side at density 353 kg/m^3 are 45 and 44 respectively. Likewise, the relationship of density (kg/m^3) and surface hardness (shore D) using water and cyclopentane as a co-blowing agent is illustrated in Figure 4 below.

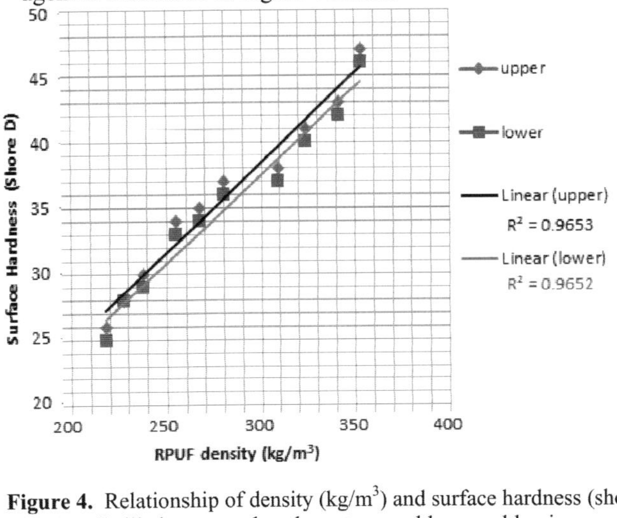

Figure 4. Relationship of density (kg/m^3) and surface hardness (shore D) using distilled water and cyclopentane co-blown as blowing agent.

Similar to full water blown, when density is increased from 218 to 353 kg/m^3, the surface hardness (shore D) is increased both on the upper and lower side of the sample. However, the surface hardness of cyclopentane co-blown (recipe 2) is higher compared to pure water blown at the same density. The average surface hardness on RPUF using CP co-blown at the upper and lower side (density 353 kg/m^3) is 47 and 46 respectively. The reason that water blown recipe has less surface hardness may be because of biuret formation at the surfaces. This white solid biuret causes brittleness to RPUF surfaces and results in lower surface hardness. The R2 value indicates that the surface hardness of RPUF is relatively depending on RPUF density for both blowing agents.

3.3 Effect of blowing agents level on glass transition temperature of rigid foams

RPUF samples in both cases of blowing at the same density were investigated for glass transition temperature (°C) by DSC analysis. The glass transition temperature of pure water blown (recipe 1) and cyclopentane co-blown (recipe 2) are shown in Table 3.

Table 3. Glass transition temperature of pure water and cyclopentane co-blown.

	Recipe 1	Recipe 2
	Glass transition temperature (°C).	
at density 353 kg/m^3	215.95	218.84

From DSC analysis result, glass transition temperature of cyclopentane co-blown is found to be higher (218.84 °C) compared to pure water blown recipe (215.95 °C). According to the network formation theory [18], the increasing of T_g illustrates higher crosslink at the surface of RPUF CP co-blown.

4 Conclusions

Water and cyclopentane (CP) co-blown can be used to produce environmental friendly rigid polyurethane foam (RPUF). The combination of both blowing agents can produce imitation wood product with desired properties, and yield proper reaction times, density (kg/m^3) and surface hardness (shore D).

The cyclopentane co-blown provides alternative reaction times to produce RPUF. The combination of exothermic heat release and heat absorption of water and CP results in the optimal time of RPUF reaction. The reaction time is longer with a higher content of CP due to heat absorption of CP.

Surface hardness of RPUF is increased when density is increased. RPUF prepared by CP co-blown recipe has a higher surface hardness (shore D: 47) compared to the full distill water blown (shore D: 45) at the same density (353 kg/m^3) due to condensed chemicals at the surfaces. Meanwhile polyuria (biuret) formed at RPUF surfaces prepared by water blown has less surface hardness.

Glass transition temperature (T_g) of the CP co-blown RPUF is slightly higher. It is increased from 215.95 °C to 218.84 °C for water and CP co-blown recipe respectively. This indicates that at the surface of CP co-blown RPUF has higher crosslink than the pure water-blown recipe. The CP co-blown RPUF creates condense chemicals at the RPUF surface. It results in higher surface hardness and higher glass transition temperature compared to pure the distill water blown RPUF.

Lastly, the obtained results of the present work show that further studies should focus on decomposition of RPUF samples using Fourier Transform Infrared (FTIR) analysis, and analysis of surface appearance via injection machine i.e., high pressure machine, and chemical test by methyl ethyl ketone (MEK), respectively.

Acknowledgments

This study was supported by Covestro, polyurethane (PUR) system house, Thailand (former Bayer materials science), providing chemicals and laboratory used. The authors would like to give a special thanks to PUR team for their kind intellectual and practical advises throughout the study.

References

1. Jiu-Jain WU and Run-Ping JIA, "Effect of Dimethyl Methylphoshonate (DMMP) flame retardant on Rigid polyurethane foam", (2017)
2. Rekha R. Rao, Lisa A. Mondy, Kevin N. Long, Mathew C. Celina, Nicholas Wyatt , et al., "The Kinetics of Polyurethane Structural Foam Formation: Foaming and Polymerization", *AIChE Journal,* Vol. 63, No. 7, July (2017)
3. Tesser R, Di Serio M, Sclafani A, Santacesaria E., "Modeling of polyurethane foam formation",. *J Appl Polym Sci.*;92:1875, (2004)
4. Dusek, K.; Spirkova, M. "Crosslinking and Network Formation in Polyurethane Systems with Excess Isocyanate"; *Cell.Polym.,* 9/2, 69- 83, (1990)
5. Wang Huia, Wang Qing-songa, He Jia-jiaa, Mao Zhan-lia, Sun Jin-huaa, "Study on the Pyrolytic Behaviors and Kinetics of Rigid Polyurethane Foams", *Procedia Engineering 52*, 377 – 385, (2013)
6. Tomy Widya and Christopher W. Macosko,"Nanoclay-Modified Rigid Polyurethane Foam"*Journal of Macromolecular Science, Part B: Physics*, 44:897–908, (2005)
7. Kraehling H; Zipfel L, " Polyurethane Foams : Environmental Competitiveness of 365-MFC-Blown Products", *Conference proceedings., Boston, Ma., 8th- 11th, p.23-31Polyurethanes Conference,* Oct. (2000)
8. Barthelemy P P; Leroy A Solvay Research & Technology,"Progress in the Development of a new HFC Blowing Agent : a closer analysis of two leading candidates", *Polyurethanes Expo '96. Conference Proceedings., Las Vegas, Nv.*, 20th-23rd, Oct. (1996)
9. W J Seo, H C Jung, W N Kim, "Mechanical, Morphological, and Thermal Properties of Rigid Polyurethane Foams Blown by Distilled Water [J]." *J. Appl. Polym. Sci.*, 90 (1): 12-21, (2003)
10. M. Thirumal,1 Dipak Khastgir,1 Nikhil K. Singha,1 B. S. Manjunath,2 Y. P. Naik, "Effect of Foam Density on the Properties of Water Blown Rigid Polyurethane Foam", *Journal of Applied Polymer Science*, Vol. 108, 1810–1817, (2008)
11. Schilling S L., "Appliance rigid foams blown with cyclopentane and cyclopentane/ isopentane blends", *J Cell Plast*, 36(3): 190-205, (2000)
12. Li Xiaobin, CAO Hongbin & ZHANG Yi. Structures and physical properties of rigid polyurethane foams with water as the sole blowing agent. *J Science in China Series B: Chemistry,* Vol.49 No.4 363-370, (2006)
13. H. C. Jung, S. C. Ryu, W. N. Kim, Y.-B. Lee, K. H. Choe , S.-B. Kim, "Properties of Rigid Polyurethane Foams Blown by HCFC 141B and Distilled Water", *Journal of Applied Polymer Science*, Vol. 81, 486–493, (2001)
14. Li Xiaobin, CAO Hongbin, ZHANG Yi, Properties of Water Blown Rigid Polyurethane, Foams with Different Functionality, *J Wuhan University of Technology-Mater. Sci. Ed.*, 2008.
15. Woods, G. The ICI Polyurethane Handbook; Wiley: New York, (1990)
16. Oertel, G. Polyurethane Handbook; Hanser: New York, (1993)
17. Klempner, D.; Frisch, K. C. Handbook of Polymeric Foams and Foam Technology; Oxford University Press: New York, (1991)
18. T Kaiser., "Highly Crosslinked Polymers[J]", *Prog. Polym. Sci.*,14 (3): 373-450, (1989)

https://doi.org/10.1051/matecconf/201818702002

The Antibacterial and Antioxidant Activity of Centella Asiatica Chloroform Extract-loaded Gelatin Nanoparticles

Kittiya Kesornbuakao[1], Suksun Amornraksa[2], Malinee Sriariyanun[2], Suvaluk Asavasanti[3] and Patchanee Yasurin[4]

[1] *Food Biotechnology Department, Biotechnology Faculty, Assumption University, Bangkok, 10240, Thailand*
[2] *The Sirindhorn International Thai-German Graduate School of Engineering, King Mongkut's University of Technology North Bangkok*
[3] *Food Technology & Engineering Laboratory, Pilot Plant Development & Training Institute, King Mongkut's University of Technology Thonburi, Bangkok, Thailand*
[4] *Food Biotechnology Department, Biotechnology Faculty, Assumption University, Bangkok, 10240, Thailand*

Abstract. Nanoencapsulation of Centella asaitica (CA) crude chloroform extract seems to be an attractive approach that may improve drug bioavailability and drug delivery system. In the study, CA extract-loaded gelatin nanoparticles (CGNP) were developed by gelatin one-step and two-step desolvation methods, using three different ratios between CA crude chloroform extract and gelatin (1:2, 1:3, and 1:4 w/w). The antibacterial and antioxidants activities of CGNP and CA were compared. The antioxidant activity of CGNP and CA was evaluated by using DPPH radical scavenging assay and ferric reducing antioxidant power (FRAP) assay. Results showed an enhancement of CGNP antibacterial activity against food borne pathogen. The highest inhibition diameters and FRAP determined from CGNP prepared by one-step at 1:4 ratios (1.03±0.39 cm and 1.23±019 mmol Fe2+/mg dried weight, respectively). However, CGNP was no significant difference in DPPH radical scavenging activity compared to CA. These results provide useful information for developing effective nanoencapsulation of CA as effective natural ingredient.

1 Introduction

Centella asiatica (CA) has been introduced into the group of medicinal plants since the ancient period. The region of CA is located in the tropical and subtropical areas mostly in South East Asia and spread through Western part of the earth. It generally consists of various chemical compositions which have ability to cure many symptoms and diseases related to bacterial infection are involved in the roots and leaves [1]. CA is known that herbs are rich in phenolic compounds [2] that have antioxidant properties as protective agent against free radical compound by scavenging or trapping methods [3, 4]. Furthermore, CA could inhibited the growth of Bacillus cereus and Listeria monocytogenes at normal, osmotic stress, and high acidic conditions, the growth of pathogenic bacteria in intestines and the growth of both Gram-positive and Gram-negative bacteria[5-8].

In fact, the bioactive compounds such as asiaticoside and asiatica acid can transfer only 50 % or even less than that into the human. That means it has a limitation on the absorption of bioactive compounds from the crude extract of CA when tested under In Vivo and acidic condition. The experiment on In Vitro gave significantly high efficiency on the drug absorption and drug delivery system [9].

Nanotechnology has been used to improve drug bioavailability and drug delivery system. Previous studies of nanoparticles, which are PGLA-nanoparticles, BSA-nanoparticles, and Gelatin-nanoparticles, indicated that nanoparticles technique is applicable to improve bioavailability [10-13]. Gelatin is defined as the hydrophilic proteins derived from collagen extracted from bones, ligaments, skin, and tendons of animal. Gelatin, known as readily and economical source, which is one of several available materials used for generating the nanoparticles, it's acts as the carrier and primary protection for the medicine to be able to increase bioavailability activity.

This research aims to compare antibacterial and antioxidants activity of CA and CA chloroform extract-loaded gelatin nanoparticle (CGNP) prepared by gelatin one-step desolvation method and two-step desolvation methods on three different ratios between CA and gelatin (1:2, 1:3, and 1:4 w/w).

2 Materials and methods

2.1 Preparation of CA

Fresh CA was purchased from Bangkapi markets in Bangkok, Thailand. The aerial part of CA was used. Fresh CA were washed with tap water and cut into small pieces. Then it was air dried in oven (Memmert UM500) at 45°C. The dried samples were finely ground into powder. The powder was kept at 4°C before used [8].

2.2 Preparation of CA crude chloroform extract

CA powder is extracted with chloroform using 1:10 ratio (g/ml). The mixtures are macerated at room temperature, 120 rpm, for 48 hours and then are filtered using whatman filter paper no.4. The CA extracts was evaporated at 45°C by rotary evaporators (BUCHI Rotavapor R-205). The crude was stored at -20°C prior to use in preparation of CGNP [8].

2.3 Preparation of Gelatin one-step desolvation nanoparticles

Gelatin was prepared under constant heat and pH at 40±1 °C, pH 3 (adjusting by 0.1 M HCl) by dissolving 600 mg of gelatin in 30 ml sterile distilled water. The gelatin nanoparticles were formed after adding CA at different ratio. The ratio between CA and gelatin were 1:2, 1:3, and 1:4 w/w. Then, adding 30 ml acetone dropwise. 100 μ L 8% v/v glutaraldehyde solution was added to stabilize of CGNP and the solution was stirred gently for 2 hours. CGNP were centrifuged and washed with distilled water. The centrifuged particles were resuspended and disperse in 3% mannitol, then freeze-dried for 48 hours. The dried CGNP were kept 4°C prior to use in antioxidant assay [13].

2.4 Preparation of Gelatin two-step desolvation nanoparticles

The encapsulation of CA in gelatin nanoparticles was modified from Azimi et al. [13]. 600 mg gelatin was added in 30 ml sterile distilled water under constant temperature at 40±1 °C. The precipitation of high molecular weight gelatin was obtained by adding 30 ml acetone into gelatin solution. After 15 minutes, the high molecular weight gelatin was redissolved with 30 ml sterile distilled water at 40±1°C, pH 3 and stirred gently. CGNP was formed by adding CA at different ratios. The ratios between CA and gelatin were 1:2, 1:3, and 1:4 w/w. Then, adding 30 ml acetone dropwise. The stabilization of CGNP was generated by adding 100 μ L glutaraldehyde solution (8 % v/v) and stirring gently for 2 hours. CGNP were centrifuged and washed with distilled water. The centrifuged particles were resuspended and disperse in 3% mannitol, then freeze-dried for 48 hours. The dried CGNP were kept 4°C prior to use in antioxidant assay.

2.5 Antimicrobial activity

CA and CGNP were added on MHA plates which were swabbed with 100 μL of bacterial cultures (approx. 1.5x108 CFU/ml) was referred to the modified agar well diffusion method adapted from Rattanakom and Yasurin [8]. Penicillin G and DMSO were used as positive and negative controls, respectively. After 24 hours of incubation, the inhibition zone were measured to determine how effective CA and CGNP could inhibit certain microorganisms which were Escherichia coli ATCC25822, Bacillus cereus, B. subtilis, Staphylococcus aureus, Salmonella enterica Typhimurium U302 (DT104b), S. enterica Enteritidis (human), and S. enterica 4,5,12:i:- (human) US clone.

2.6 Antioxidant activity by Ferric reducing antioxidant power (FRAP) assay

The modified ferric reducing antioxidant potential assay [14] was used to determine FRAP value of CA and CGNP. The FRAP reagent was prepared using 300 mmol sodium acetate buffer at pH 3.6, 20 mmol iron chloride and 10 mmol 2,4,6-tripyridyl-s-triazine dissolved in 40 mmol hydrochloric acid at a ratio of 10:1:1 (v:v:v).The reagent was incubated at 37° C for 10 minutes before use. The 20 μ l of 1 mg/ml of CA and CGNP was added, followed by adding 1000 μ l of FRAP reagent vigorously and kept in the dark for 30 minutes. The optical density (OD) of this mixture was measured at 593 nm. FRAP values were expressed as mmol Fe2+/mg of sample. All measurements were done in triplicate and three replications independently.

2.7 Antioxidant activity by 2,2-Diphenyl-1-picrylhydrazil (DPPH) radical scavenging activity

This method is modified from Brand-Williams [15] to determine the antioxidant activities of CA and CGNP. The mixture of 100 μ L CA or CGNP and 3.9 mL of methanol DPPH solution (5 x 10-5 mol/L) is shaken thoroughly and kept in the dark room for 30 minutes. The mixture is measured by UV-vis spectrophotometer at 517 nm. The unit of μ g/mL of gallic acid equivalent (GAE) per 1 g sample is used to express results.

2.8 Statistical analysis

All experiments were conducted in three replications and statistical analysis was accomplished using ANOVA with Duncan's multiple range tests ($p < 0.05$) by SAS software version 9.4.

3 Results and discussion

3.1 Antioxidant activities

As the antioxidant activity of CA come from the complex of phytochemicals and involvement of many reactions, the measurement of the activity cannot be done accurately with only one assay. The combination of DPPH radical scavenging activity and Ferric reducing antioxidant power (FRAP) were used to evaluate the antioxidant activity.

The antioxidant activity of all CGNPs and CA were statistically analyzed to see the different between each sample that were prepared from different preparation methods and ratios. The results from Figure 1 and Figure 2 demonstrated that CGNP prepared by one-step desolvations method showed significantly greater FRAP than both CA and CGNP two-step at all ratios ($p<0.05$).

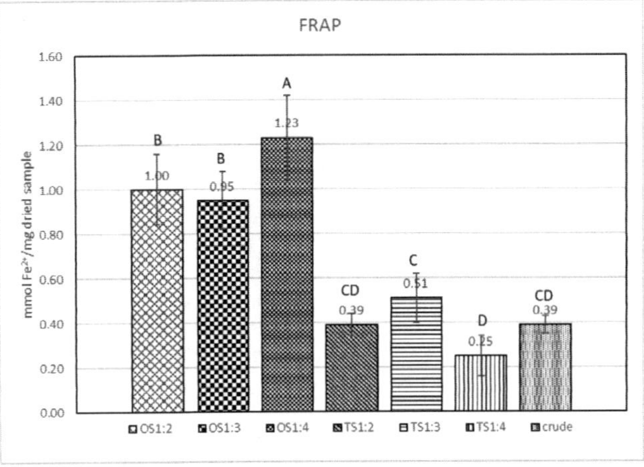

Figure 1. Antioxidant activity by Ferric reducing antioxidant power (FRAP) assay of CA and CGNP prepared at different ratios between CA and Gelatin. OS and TS represented CGNP one-step and CGNP two-step respectively. A, B represented significantly different value in a column at $p<0.05$.

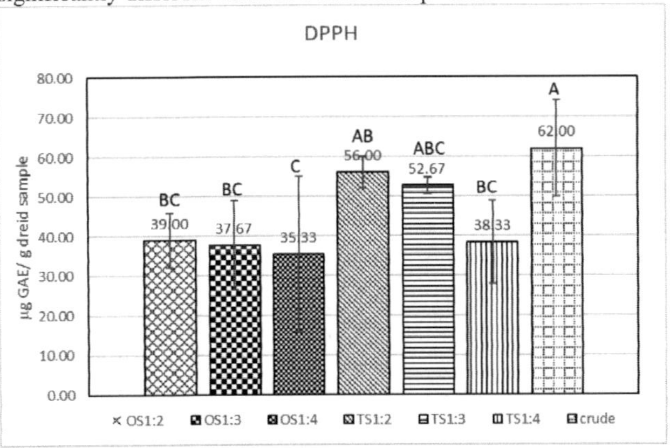

Figure 2. Antioxidant activity by 2, 2-Diphenyl-1-picrylhydrazil (DPPH) radical scavenging activity of CA and CGNP prepared at different ratios between CA and Gelatin. OS and TS represented CGNP one-step and CGNP two-step respectively. A, B represented significantly different value in a column at $p<0.05$.

According to FRAP assay, determination of ability to reduce of ferric iron (Fe^{3+}) to ferrous iron (Fe^{2+}) [16], CA nanoencapsulated could enhance FRAP of CA. In addition, CGNP one-step nanoencapsulation structure may differ from two-step due to size of the protein and effect to FRAP values by releasing of CA. Moreover, CGNP prepared by one-step desolvation method at ratio 1:4 showed highest FRAP (1.23±019 mmol Fe^{2+}/mg dried weight) among CGNP. However, CGNP one-step at all ratios exhibited significantly lower in DPPH radical scavenging activity than CA ($p<0.05$). DPPH radical scavenging assay measures the reducing ability of antioxidants toward DPPH [16]. The decrease in antioxidant activity of nanoencapsulated active compounds is probably related to the protectability of the droplets in the system that slow down the reaction rate of reduction [17]. In contrast, CGNP prepared with two-step desolvation method at ratio 1:2 and 1:3 showed no significantly different in DPPH radical scavenging activity compared with CA (56.00 ± 4.00 and 52.67 ± 2.08 mmol Fe^{2+}/mg dried weight, respectively).

3.2 Antibacterials

CGNPs were prepare by one-step and two-step gelatin desolvation methods[13]. Both CGNP preparation were done with different ratio of CA to gelatin at 1:2, 1:3, and 1:4. The well agar diffusion method was used for evaluating antibacterial activity of CGNPs and CA with different concentration (100, 200, and 300 µl/ml) against seven food borne pathogens (*Escherichia coli* ATCC25822, *Streptococcus aureus, Bacillus cereus, B. subtilis, Salmonella enterica* Typhimurium U302 (DT104b), *S. enterica* Enteritidis (human), and *S. enterica* 4,5,12:i:- (human) US clone).

The results showed that antibacterial activity of CGNP were clearly increased on *S.s aureus, S. enterica* Typhimurium U302 (DT104b), *S. enterica* Enteritidis (human), and *S. enterica* 4,5,12:i:- (human) US clone comparing with CA as shown in Table 1. The CGNP likely more affected to gram-negative (*E. coli* ATCC25822, *S. enterica* Typhimurium U302 (DT104b), *S. enterica* Enteritidis (human), and *S. enterica* 4, 5, 12: i: - (human) US clone) than gram-positive bacteria (*B. cereus, B. subtilis*, and *S. aureus*). Because the structure of gram-negative and gram-positive bacteria are different. Gram-negative bacteria has inner membrane, thin layer made of peptidoglycan, and outer membrane whereas gram-positive has an inner membrane covered with thick layer of peptidoglycan [18, 19]. The thick layer of peptidoglycan have high protectaility for gram-positive bacteria which harder for CGNP to get through the cells. In addition, the highest inhibition zone of CGNP was 1.03±0.39 cm against *S. enterica* Enteritidis using gelatin one-step desolvation method on ratio 1:4, 300 µg/ml.

4 Conclusion

Overall, CGNPs exhibited interesting antibacterial and antioxidant activity. The DPPH radical scavenging of CGNP tend to decrease because it could protect an active compound and slow down the oxidative reaction. CGNPs enhance antibacterial activity on human pathogen, especially on gram-negative bacteria. These results provide a promising for improving bioavailability of CA

Acknowledgements

This work was supported by the research grant of Research and Researchers for Industries (RRI) by Thailand Research Fund (TRF) (PHD59I0074) and King Mongkut's University of Technology, North Bangkok

(Research University Grant No. KMUTNB-61-GOV-01-55).

References

1. A. J. Das, J. Biol. Act. Prod. Nat., **1:4** (2011)
2. P. Yasurin, M. Sriariyanun, T. Phusantisampan, KMUTNB Int. J. Appl. Sci. Technol., **9** (2016)
3. D de Beer, E. Jouber, W.C.A. Gelderblom, and M. Manley, South African Journal of Enology & Viticulture, **23** (2002)
4. D. Huang, B. Ou, and R. L.Prior, J. Agric. Food Chem., **53** (2005)
5. N. Pitinidhipat and P. Yasurin, AU J.T., **15** (2012)
6. C. V. Utami, N. Pitinidhipat, and P. Yasurin, KMITL Sci. Tech. J., **12** (2012)
7. B. Mamtha, K. Kavitha, K. K. Srinivasan, and P. G. Shivananda, Indian J. Pharmacol, **36** (2004)
8. S. Rattanakom and P. Yasurin, Orient. J. Chem., **34** (2015)
9. W. Kriengsinyos, C. Chitchumroonchokchai, and P. Siriprapa, FASEB Journal, **20** (2006)
10. X. Xie, Q. Tao, Y. Zou, F. Zhang, M. Guo, Y. Wang, H. Wang, Q. Zhou, and S. Yu, J. Agric. Food. Chem., **59** (2011)
11. Y. Zhan, Y. Min, Z. Zhang, G. Hong, and Q. Xiong, anoscale Res. Lett., **9** (2014)
12. K. Kesornbuakao, and P. Yasurin, Oreintal Journal of Chemistry., **32** (2016)
13. B. Azimi, P. Nourpanah, M. Rabiee, and S. Arbab, Iran. Biom. J., **18** (2014)
14. IF. Benzie, and JJ. Strain, Methods Enzymol., **299** (1999)
15. W. Brand-Williams, M. E. Cuvelier, and C. Berset, LWT - Food Science and Technology, **28** (1995)
16. K.D.R.R. Silva, and M.S.F. Sirasa, Food Chemistry, **238** (2016)
17. M. B. Jemaa, H. Falleh, R. Serairi, M. A. Neves, M. Snoussi, H. Isoda, M. Nakajima, and R. Ksouri, Innovative Food Science & Emerging Technologies., **45** (2018)
18. M. Madigan and J. Martinko, 11th ed., Englewood Cliffs. NJ: Prentice Hall, (2005)
19. L. Brown, J.M. Wolf, R. Prados-Rosales, and A. Casadevall, Nat. Rev. Microbiol., **13** (2015)

Table 1. The inhibition zone of CGNP and crude extract against 7 different microorganisms in the unit of centimeter.
Note: Superscript in capital letters (A, B, C) and small letters (a, b, c) represented significantly different value in a column and a row at $p<0.05$, respectively ST stands for S. enterica Typhimurium, SE stands for S. enterica Enteritidis, and SUS stands for S. enterica 4,5,12:i:- (human) US clone.

Sample	Ratio (crude:gelatin)	Concentration (ug/ml)	Inhibition zone (cm.)						
			E. coli	*S. aureus*	*B. cereus*	*B. subtilis*	*ST*	*SE*	*SUS*
Gelatin one-step desovation nanoparticles	1:2	100	$0.74\pm0.18^{BC,ab}$	$0.68\pm0.19^{B,ab}$	$0.86\pm0.11^{BC,a}$	$0.73\pm0.12^{BCD,ab}$	$0.68\pm0.12^{B,ab}$	$0.61\pm0.17^{CDEFG,b}$	$0.66\pm0.14^{CDEF,ab}$
		200	$0.87\pm0.16^{BC,a}$	$0.60\pm0.10^{B,b}$	$0.84\pm0.11^{BCD,a}$	$0.84\pm0.11^{B,a}$	$0.69\pm0.05^{B,ab}$	$0.77\pm0.17^{BCDE,ab}$	$0.73\pm0.18^{BCDE,ab}$
		300	$0.84\pm0.03^{BC,a}$	$0.69\pm0.25^{B,a}$	$0.92\pm0.19^{B,a}$	$0.88\pm0.19^{B,a}$	$0.81\pm0.13^{B,a}$	$0.89\pm0.14^{BC,a}$	$0.79\pm0.19^{BCD,a}$
	1:3	100	$0.80\pm0.11^{BC,a}$	$0.80\pm0.11^{B,a}$	$0.72\pm0.26^{CDE,a}$	$0.76\pm0.15^{BCD,a}$	$0.63\pm0.21^{B,a}$	$0.62\pm0.17^{CDEFG,a}$	$0.62\pm0.18^{DEF,a}$
		200	$0.86\pm0.10^{BC,a}$	$0.75\pm0.13^{B,a}$	$0.72\pm0.18^{CDE,a}$	$0.79\pm0.18^{BC,a}$	$0.69\pm0.18^{B,a}$	$0.54\pm0.33^{DEFG,a}$	$0.67\pm0.16^{BCDEF,a}$
		300	$0.85\pm0.06^{BC,a}$	$0.76\pm0.12^{B,a}$	$0.73\pm0.13^{CDE,a}$	$0.80\pm0.24^{BC,a}$	$0.81\pm0.12^{B,a}$	$0.69\pm0.19^{CDEF,a}$	$0.79\pm0.11^{BCD,a}$
	1:4	100	$0.82\pm0.12^{BC,a}$	$0.69\pm0.17^{B,a}$	$0.67\pm0.27^{CDE,a}$	$0.65\pm0.17^{BCD,a}$	$0.61\pm0.18^{B,a}$	$0.61\pm0.20^{CDEFG,a}$	$0.72\pm0.08^{BCDE,a}$
		200	$0.87\pm0.10^{BC,a}$	$0.71\pm0.19^{B,a}$	$0.71\pm0.20^{CDE,a}$	$0.84\pm0.24^{B,a}$	$0.77\pm0.11^{B,a}$	$0.79\pm0.11^{BCD,a}$	$0.75\pm0.08^{BCD,a}$
		300	$0.80\pm0.14^{BC,a}$	$0.72\pm0.15^{B,a}$	$0.74\pm0.21^{BCDE,a}$	$0.85\pm0.19^{B,a}$	$0.72\pm0.09^{B,a}$	$1.03\pm0.39^{B,a}$	$0.82\pm0.04^{BC,a}$
Gelatin two-step desovation nanoparticles	1:2	100	$0.82\pm0.07^{BC,a}$	$0.67\pm0.03^{B,a}$	$0.67\pm0.15^{CDE,a}$	$0.67\pm0.16^{BCD,a}$	$0.64\pm0.14^{B,a}$	$0.68\pm0.25^{CDEF,a}$	$0.75\pm0.13^{BCD,a}$
		200	$0.84\pm0.14^{BC,a}$	$0.69\pm0.08^{B,a}$	$0.72\pm0.05^{CDE,a}$	$0.72\pm0.16^{BCD,a}$	$0.63\pm0.14^{B,a}$	$0.80\pm0.19^{BCD,a}$	$0.78\pm0.18^{BCD,a}$
		300	$0.91\pm0.11^{B,a}$	$0.69\pm0.06^{B,a}$	$0.67\pm0.10^{CDE,a}$	$0.86\pm0.24^{B,a}$	$0.68\pm0.09^{B,a}$	$0.88\pm0.09^{BC,a}$	$0.86\pm0.26^{B,a}$
	1:3	100	$0.71\pm0.12^{BC,a}$	$0.66\pm0.06^{B,a}$	$0.65\pm0.18^{DE,a}$	$0.74\pm0.09^{BCD,a}$	$0.69\pm0.18^{B,a}$	$0.68\pm0.26^{CDEF,a}$	$0.68\pm0.19^{BCDEF,a}$
		200	$0.79\pm0.12^{BC,a}$	$0.70\pm0.07^{B,a}$	$0.61\pm0.11^{E,a}$	$0.76\pm0.24^{BCD,a}$	$0.72\pm0.07^{B,a}$	$0.62\pm0.19^{CDEFG,a}$	$0.73\pm0.08^{BCDE,a}$
		300	$0.88\pm0.10^{BC,a}$	$0.73\pm0.03^{B,a}$	$0.72\pm0.16^{CDE,a}$	$0.75\pm0.15^{BCD,a}$	$0.63\pm0.10^{B,a}$	$0.75\pm0.31^{CDE,a}$	$0.80\pm0.15^{BCD,a}$
	1:4	100	$0.85\pm0.12^{BC,a}$	$0.70\pm0.11^{B,a}$	$0.66\pm0.20^{DE,a}$	$0.70\pm0.08^{BCD,a}$	$0.68\pm0.06^{B,a}$	$0.64\pm0.19^{CDEFG,a}$	$0.75\pm0.18^{BCD,a}$
		200	$0.80\pm0.14^{BC,a}$	$0.68\pm0.08^{B,ab}$	$0.64\pm0.14^{E,b}$	$0.75\pm0.13^{BCD,ab}$	$0.72\pm0.06^{B,ab}$	$0.75\pm0.10^{CDE,ab}$	$0.75\pm0.06^{BCD,ab}$
		300	$0.86\pm0.12^{BC,a}$	$0.71\pm0.10^{B,a}$	$0.69\pm0.14^{CDE,a}$	$0.81\pm0.22^{BC,a}$	$0.72\pm0.02^{B,a}$	$0.69\pm0.07^{CDEF,a}$	$0.79\pm0.12^{BCD,a}$
Crude		100	$0.80\pm0.14^{BC,a}$	$0.64\pm0.10^{B,ab}$	$0.70\pm0.19^{CDE,ab}$	$0.66\pm0.19^{BCD,ab}$	$0.62\pm0.03^{B,ab}$	$0.39\pm0.13^{G,c}$	$0.53\pm0.03^{F,bc}$
		200	$0.69\pm0.45^{C,a}$	$0.58\pm0.22^{B,a}$	$0.69\pm0.10^{CDE,a}$	$0.57\pm0.20^{CD,a}$	$0.47\pm0.15^{B,a}$	$0.42\pm0.03^{FG,a}$	$0.55\pm0.08^{EF,a}$
		300	$0.77\pm0.17^{BC,a}$	$0.46\pm0.40^{B,a}$	$0.74\pm0.12^{BCDE,a}$	$0.52\pm0.20^{D,a}$	$0.50\pm0.15^{B,a}$	$0.49\pm0.05^{EFG,a}$	$0.54\pm0.07^{EF,a}$
PennicilinG (positive control)		300	$1.11\pm0.06^{A,c}$	$3.25\pm0.89^{A,a}$	$1.15\pm0.30^{A,c}$	$2.43\pm0.35^{A,ab}$	$2.09\pm0.69^{A,b}$	$1.90\pm0.39^{A,bc}$	$2.11\pm0.08^{A,b}$

[a] Corresponding author: patchaneeYsr@au.edu

DMSO (negative control)			$0.00\pm0.00^{D,a}$	$0.00\pm0.00^{C,a}$	$0.00\pm0.00^{F,a}$	$0.00\pm0.00^{E,a}$	$0.00\pm0.00^{C,a}$	$0.00\pm0.00^{H,a}$	$0.00\pm0.00^{G,a}$

Preparation of Microcrystalline Cellulose from Water Hyacinth Reinforced Polylactic Acid Biocomposite

Teerapa Semachai[1], Panitnad Chandranupap[1] and Pravitra Chandranupap[2]

[1] *Department of Chemical Engineering, Faculty of Engineering, King Mongkut's University of Technology North Bangkok, Thailand*
[2] *Department of Industrial Chemistry, Faculty of Applied Science, King Mongkut's University of Technology North Bangkok, Thailand*

Abstract. In this work, we successfully mixed polylactic acid (PLA) with microcrystalline cellulose (MCC) from water hyacinth. The MCC was prepared by treating water hyacinth fiber (WHF). Then hydrochloric acid was used to hydrolyze treated fiber to MCC. X-rays diffraction (XRD) showed that the MCC produced has 73.28 per cent crystallinity. Internal mixing was used to combine composites between MCC and PLA. Percentages of MCC were 1, 5, 10 and 15, respectively. Fourier transform infrared (FT-IR) spectroscopy indicated that the interaction between MCC and PLA are only mechanically interaction. Tensile testing of this composite (ASTM D638) revealed that tensile strength and percentage of elongation at break decreased but the increase of young's modulus. The morphological analysis was observed thru composites fractured surface by Scanning Electron Microscope (SEM). They showed a void between cellulosic fiber and PLA when high amount of MCC conformed with tensile results.

1 Introduction

Nowadays, there is an increasing demand for products made from renewable and sustainable non-petroleum based resources. Biopolymer-based materials have been interested because they are environmental friendly materials [1].

Polylactic acid (PLA) is the second most important bioplastic of the world in regard to consumption in volume [2]. It has been used in several applications such as food packaging, water and milk bottles, barriers for sanitary products and diapers, as well as in automotive applications [3]. Its can be derived from renewable resources like starch so it has high water vapor permeability but poor mechanical properties. To improve such properties, some additives or other ingredients such as nanoclays, fibers, carboxymathylcellulose (CMC), microcrystalline cellulose (MCC), etc. have been added [4].

Microcrystalline cellulose (MCC) is native from cellulose, which can be found in wood pulp. The cellulose polymers in pulp are reduced to small chain polymers or microcrystals by hydrolysis. Their application in composite materials has attracted attention due to their high mechanical strength, high stiffness, renewability, biodegradability and low weight. MCC is characterized by a high degree of crystallinity, and the values typically range between 55% and 80% [5]. Lignocellulosic material was generally used to produce MCC such as kenaf fiber, cotton, oil palm biomass, etc.

Water hyacinth is an aquatic plant which can live and reproduce floating freely. It became dense coverage that prevents the sunlight from penetrating the water due to very fast-growing plant. The weed removes oxygen from the water, resulting in the death of many fish. Many countries, including Thailand, try to convert this plant into valuable things, such as material for papermaking, charcoal briquetting and etc. [6]. The aim of this research is to prepare MCC from water hyacinth and study the effect on morphological, thermal and mechanical properties of PLA-MCC composite.

2 Experimental

2.1 Materials

Polylactic acid (Nature Work TM PLA 2003D) was purchased from BC polymer marketing Co. Ltd. It has a specific gravity of 1.24 g/cm3 and melt flow index (MFI) of 6 g/10min (210°C, 2.16kg). Water hyacinth fiber was purchased from Water hyacinth handicraft products group (Phayao, Thailand).

2.2 Preparation of MCC

Water hyacinth fiber (WHF) was soaked in sodium hydroxide solution (10% w/v and 15% w/v) for 1 hour at 60°C then washed until neutral and dried it in oven. 2M Hydrochloric acid was used to treat fiber for 6 hours at

80°C. Hydrolyzed fiber was then washed and sieved to collect size below 100 μm. The hydrolyzed fiber was coded as MCC10 (or MCC15, depended on NaOH condition).

2.3 Preparation of PLA and PLA/MCC composite

About 59.4 g (57 g, 54 g and 51 g) of PLA pellets were mixed in an internal mixer for 5 min, then 0.6 g (3 g, 6 g and 9 g) of MCC were added. PLA and MCC were blended at 175°C until 15 min of process, mechanically stirred up to 38 rpm. The obtained mixture was coded as PLA/MCC1 (PLA/MCC5, PLA/MCC10 and PLA/MCC 15).

2.4 Sample preparation

Around 30 g of neat PLA and PLA composites were compression molded in a heated press at 180°C for 5 min without pressure. Then a pressure of 400 bar was applied for 5 min. After this period, the pressure was release and the press plates were cooled below 60°C by cold water. Finally, the mold was removed from plates and sample sheet (1.2 mm thickness) was obtained.

2.5 Characterization

2.5.1 X-ray diffraction (XRD)

The XRD data were collected using an X-Ray diffractrometer (BRUKER model D8 Advance, Germany) equipped with Cu radiation at 40 kV and 30 mA to investigate the XRD spectra of the cellulosic sample. Scattered radiation was detected in the range 2θ = 5-80°. A scanning step of 0.02° is used at a time of 1 sec per step.

2.5.2 Differential scanning calorimetry (DSC)

Differential scanning calorimeter (Mettler Toledo TGA/DSC 1, USA) was used to study thermal behavior of PLA composite. The samples were heated from 30°C to 190°C at a heating rate of 10°C/min under nitrogen atmosphere.

2.5.3 Fourier-transform infrared (FT-IR) spectroscopy

The FT-IR spectra of the cellulosic samples were recorded using a FT-IR Spectrometer (Perkin Elmer Spectrum 2000, USA) with a resolution of 4 cm^{-1} and within the wave number range 4000-400 cm^{-1}.

2.5.4 Tensile testing

The test specimens of sample for tensile according to ASTM D638 type 5. Dog-bone shaped composite samples were cut by using a computer numerical control (CNC) machine. The tensile tests were operated using a universal testing machine (Testometric M500-25AT, UK), with a 10 kN load cell with 1 mm/min crosshead speed. At least five samples were tested for each material and average values are presented.

2.5.5 Scanning electron microscope (SEM)

The Scanning Electron Microscope (SEM) was used to study morphology of MCC composites. Samples were sputter-coated with gold (Sputter coater, SPI Supplies, USA) to enhance the electrical conductivity. SEM was conducted using a JEOL JSM-6480LV model, a magnification of 300× and 1000× were applied.

3 Results & discussion

3.1. X-ray diffraction

The crystallinity of MCC has been analyzed by x-ray diffractometer. Figure 1 showed that both MCC10 and MCC15 diffractograms showed only one outstanding peak around 2θ = 22.5° (for 2 0 0 plane). That is the characteristic of cellulose.

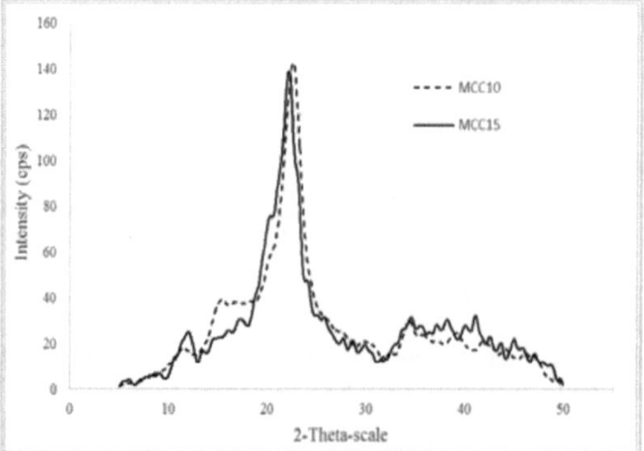

Figure 1. XRD of WHF treated with sodium hydroxide and hydrochloric acid.

The crystallinity (I_C) was determined by means of equation (1) using the height of 2 0 0 peak (I_{200}, 2θ = 22.5°) and the minimum between the 2 0 0 and 1 1 0 peaks (I_{am}, 2θ = 18°). I_{200} stands for both crystalline and amorphous material while I_{am} stands for amorphous material [7].

$$I_C = \left[\frac{I_{200} - I_{am}}{I_{200}}\right] \times 100 \quad (1)$$

The synthesis cellulosic has I_C around 73.28% and 76.44% for MCC10 and MCC15. The high percentage of crystallinity in confirmed that they are MCC.

3.2 Differential scanning calorimetry

Thermograph from DSC displays in Figure 2, neat PLA has a single melt peak, however with the addition of MCC showed a double peak forms. This double peak can be attributed to the coexistence of two crystalline populations or re-crystallization [8].

Table 1 showed the thermal properties of MCC composite glass temperature, crystallization temperature and melting temperature. It was remarkable to note that crystallization temperatures are decrease compared to neat PLA with the MCC addition into polymer matrix. Suggesting that the amount of MCC did not affect to glass and melting temperature of composites. Therefore it cannot be a nucleating agent due to less amount of MCC.

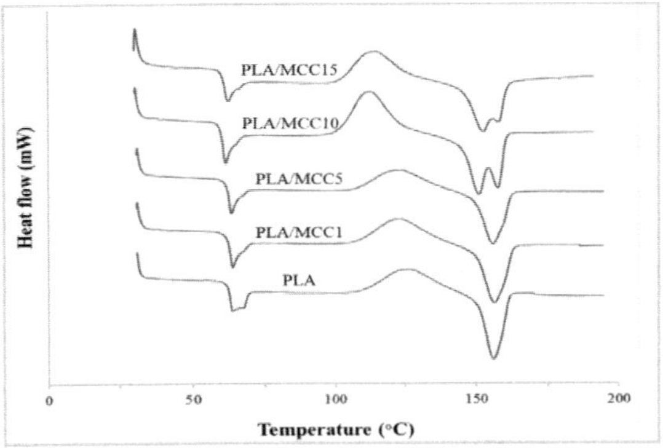

Figure 2. DSC thermographs of neat PLA and PLA/modified cellulose composites.

Table 1. Results from DSC.

	T_g (°C)	T_{cc} (°C)	T_m (°C)
PLA	59.46	121.54	150.35
PLA/MCC1	60.54	118.39	150.65
PLA/MCC5	59.07	118.53	150.00
PLA/MCC10	59.20	109.98	148.77
PLA/MCC15	59.97	111.89	149.07

3.3 Fourier transforms infrared spectroscopy

The FT-IR analysis has been widely used to identify the interaction and phase behavior of polymer composites [10]. The FT-IR spectra of PLA and PLA/MCC composites were shown in Figure 3. Figure 3a exhibited characteristic spectrum of neat PLA. The broad band at 3330–3600 cm^{-1} is O-H bending and starching for PLA. The peaks originated from the asymmetric stretching vibration of C-H are situated at 2880 to 2999 cm^{-1}.

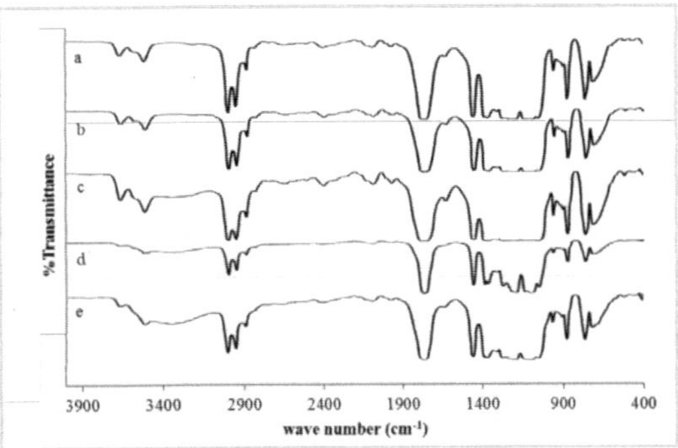

Figure 3. FT-IR spectra of PLA and PLA/MCC composites. They display (a) neat PLA (b) PLA/MCC1 (c) PLA/MCC5 (d) PLA/MCC10 and (e) PLA/MCC15.

The C-O stretching vibration of carbonyl is apparent at 1760 cm^{-1} and the peaks corresponding to C-H deformation in the CH_3 appeared at 1450 cm^{-1} [11].

Figure 3b-3e showed the character peaks of PLA/MCC composites. They did not present any new peak compared to neat PLA. This result suggested that MCC and PLA composites were produced from physically interaction rather than chemical interaction [9].

3.4 Tensile properties

The trend of tensile properties of PLA and PLA/ MCC composites was shown in Figure 4.

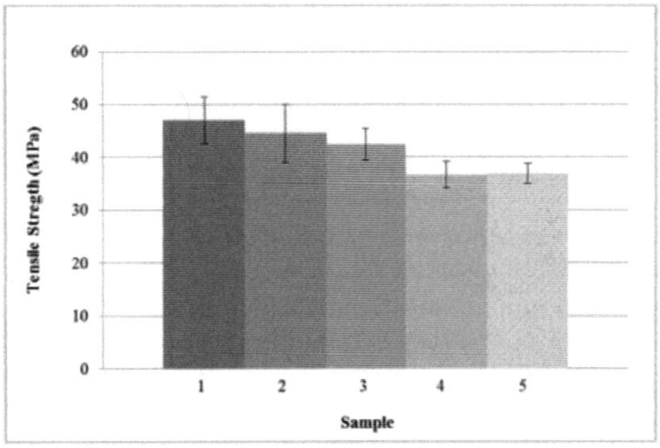

(a) Tensile strength

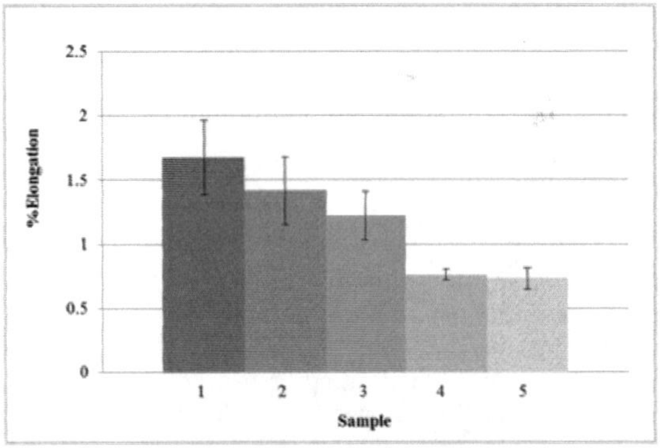

(b) Percentage of Elongation

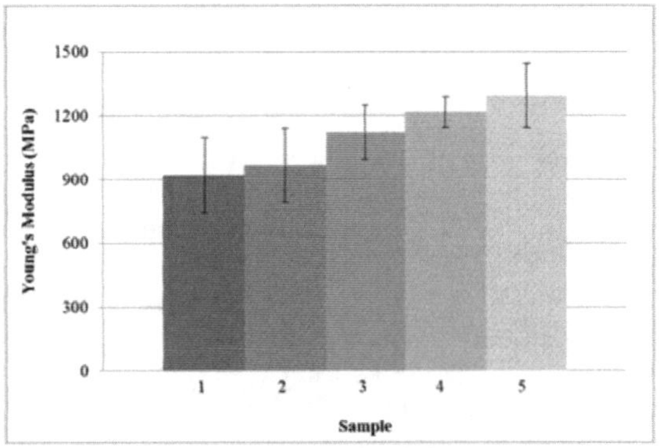

(c) Young's Modulus

Figure 4. Tensile properties of PLA and PLA/MCC composites include (a) Tensile strength (b) Percentage of Elongation and (c) Young's Modulus, the samples are (1) neat PLA (2) PLA/MCC1 (3) PLA/MCC5 (4) PLA/MCC10 and (5) PLA/MCC15.

From the result in Figure 6, addition of MCC into PLA matrix did not reveal any improvements in tensile strength and elongation at break when compared to neat PLA. In the other hand, the Young's modulus increased from 919 To 1293 MPa with the increase of MCC loading. This is due to more hydrogen bonding, stiffening effect and high crystallinity index of the filler which is a characteristic of filler/polymer composite [11]. The aggregation of MCC may cause low tensile strength and elongation at break for composites due to Van der Waal's forces [12]. Because of this phenomenon, the filler-filler interaction becomes more effect than filler-matrix interaction. So they are poor interfacial adhesion between MCC and matrix, numerous voids can be generated at filler matrix interface as a result in worse tensile properties than neat PLA.

3.4 Scanning electron microscope

The cross-sectional surface of composites was used to observe distribution of MCC and study the interaction between two components. Small amount of cellulose was distributed into PLA matrix can be seen in Figure 5a-5b. They had a rougher surface compared to high percentage of MCC (Figure 5c-5d).

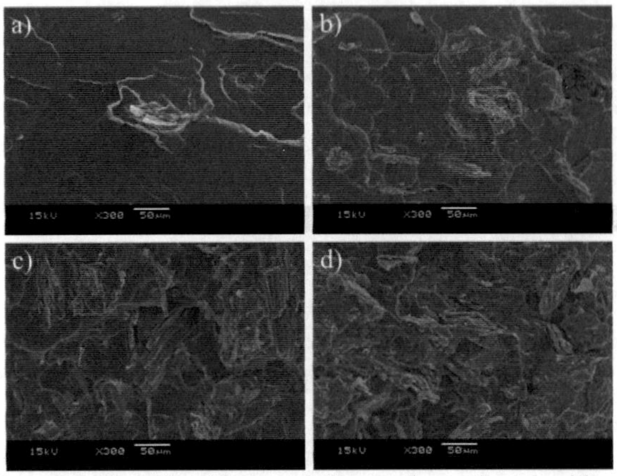

Figure 5. Scanning Electron Micrograph conducted with a magnification 300×. They exhibit (a) PLA/MCC1 (b) PLA/MCC5 (c) PLA/MCC10 and (d) PLA/MCC15.

MCC agglomeration appeared to occur with increased filler. More magnifying power demonstrated that they protruded out of fractured surface as shown in Figure 6c-6d and also visible a void between cellulosic fiber and PLA. For these reasons indicate that the dispersion of MCC was not uniform and presented poor interfacial adhesion between filler and matrix [2, 9, and 11]. These can be the explanation for the tensile results that tensile strength and elongation at break decreased when increasing amount of MCC.

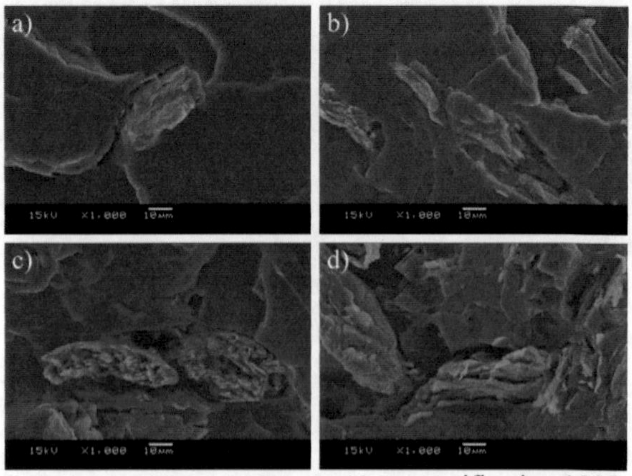

Figure 6. SEM image conducted with a magnification 1000× pictured of (a) PLA/MCC1 (b) PLA/ MCC5 (c) PLA/MCC10 and (d) PLA /MCC15.

4 Conclusions

MCC was successfully produced from water hyacinth and also mixed with PLA by internal mixing machine. The FT-IR spectra of PLA/MCC composites show no significant changes in the peaks position suggesting that only physical interaction between PLA and MCC occurred. Tensile properties demonstrate increasing of Young's modulus due to increase in hydrogen bonding and stiffening effect. In the other hand both of tensile strength and elongation at break decreased with increasing MCC loading because of a void between MCC and PLA resulted to poor interfacial adhesion. The mechanical properties may be further improved by using suitable coupling agents. This is expected to better tensile properties of composites.

References

1. M. Avella, G. B. Gaceva, A. Bužarovska, M. E. Errico, G. Gentile, A. Grozdanov. Poly (lactic acid)-Based Biocomposites Reinforced with Kenaf Fibers. *Journal of Applied Polymer science.* **108**, 3542–3551 (2008)
2. S. Joshi, U. Sharma, G. Goswami. Bio-Plastic from Waste Newspaper. *International Conference on Emerging Trends of Research in Applied Sciences and Computational Techniques* (2014)
3. M. Jonoobi, J. Harun, A. P.Mathew, K. Oksman. Mechanical properties of cellulose nano-fiber (CNF) reinforced polylactic acid (PLA) prepared by twin screw extrusion. *Composites Science and Technology.* **70**, 1742-1747 (2010)
4. Y. Byun, Y. T. Kim. Chapter 14 – Bioplastics for Food Packaging: Chemistry and Physics. *Innovations in Food Packaging (Second Edition)* 353–368 (2014)
5. D. Trache, M. H. Hussin, C. T. Hui Chuin, S. Sabar, M.R. Fazita, O. F. Taiwo, T.M. Hassan, M.K. M. Haafiz. Microcrystalline cellulose: Isolation, characterization and biocomposites application A review. *International Journal of Biological Macromolecules* (ISSN: 0141-8130, 2016)

6. R. Teygeler. Water hyacinth paper. Contribution to a sustainable future. *Papier en Water/Paper and Water*. 168-188 (2000)
7. A. Mandal, D. Chakrabarty. Isolation of nano-cellulose from waste sugarcane bagasse (SCB) and its characterization. *Carbohydrate Polymers*. **86**, 1291– 1299 (2008)
8. A. Abdulkhani, J. Hosseinzadeh, A. Ashori, S. Dadashi, Z. Takzare. Preparation and characterization of modified cellulose nanofibers reinforced polylactic acid nanocomposite. *Polymer Testing*. **35**, 73-79 (2014)
9. S. Ummartyotin, C. Pechyen. Microcrystalline cellulose and polypropylene based composite: A simple, selective and effective material for microwavable packaging. *Carbohydrate Polymers*. **142**, 133-140 (2016)
10. T. Kemala, E. Budianto, B. Soegiyono. Preparation and characterization of microspheres based on blend of poly (lactic acid) and poly (ε-caprolactone) with poly (vinyl alcohol) as emulsifier. *Arabian Journal of Chemistry*. **5**, 103–108 (2012)
11. M.K. Mohamad Haafiza, A. Hassana, Z. Zakariac, I.M. Inuwaa, M.S. Islamd, M. Jawaide. Properties of polylactic acid composites reinforced with oil palm biomass microcrystalline cellulose. *Carbohydrate Polymers*. **98**, 139-145 (2013)
12. G. H. Yew, A. M. Mohd Yusof, Z. A. Mohd Ishak, U. S. Ishiaku. Water absorption and enzymatic degradation of poly (lactic acid)/rice starch composites. *Polymer Degradation and Stability*. **90**, 488–500 (2005)

Wave Absorption and Thermal Stability of Nickel Coated Al_2O_3 Core-shell Composite Powder

Hu We[1]i, Li Cheng[1]

[1]*College of Materials Science &Technology, Nanjing University of Aeronautics and Astronautics, Nanjing 210016, China*

Abstract. As a composite cermet material, the nickel coating on the surface of alumina particles has important application value because of its good electromagnetic wave absorption performance. In this study, nickel was deposited on the surface of alumina nanopowders by hydrothermal method and sintering method. The results show that the XRD results of nickel alumina show that Ni-Al_2O_3 nanoparticles has FCC face-centered cubic structure. The grain size is about 50 nm. The absorption characteristics of electromagnetic waves were measured at 2-18 GHz by HP 8722 es microwave network analyzer. The results show that Ni-Al_2O_3 nanoparticles has certain wave-absorbing properties.

1 Introduction

The surface effect, the volume effect and the quantum size effect of nano-magnetic metal powder make it produce many physical and chemical effects different from the bulk material, and have both free electron absorption and magnetic loss, and have excellent absorption properties for high frequency electromagnetic wave[1, 2]. This is because, compared with the traditional particles, nano-particles are easy to generate interfacial polarization and multiple scattering or reflection of electromagnetic waves, because of their small particle size, large specific surface area, high surface atomic ratio and increased dangling bonds, which make them as an important wave absorption mechanism. At the same time, the quantum size effect of nano-materials causes split of the electron energy level of nano-particles, and the split energy level interval is just in the energy range of microwave, which leads to the formation of new wave absorption channels [3-5]. Various metallic magnetic material systems with nanoflake morphology have been studied to enhance high-frequency permeability [6 - 8].

However, the high permittivity and the easy oxidation and corrosion of metallic magnetic nanoflakes inhibit the application in practical microwave absorption. In particular, the high permittivity damages impedance match that is one of the keys in microwave absorption [9]. In this work, considering that Ni-Al_20_3 has high resistivity and long-term resistance to oxidation and corrosion, the Ni-Al_20_3 nanoflakes were coated with thin Ni-Al_20_3 by a simple process to improve the microwave absorption performance of Ni-Al_20_3 nanoflakes. The morphology, micro- structure, phase structure, and static magnetic properties were also investigated in detail. Among them, alumina is used as a group because of its good thermal stability and mechanical properties, and the powder coating obtained by hydrothermal method is simple and easy to obtain, which can be used as a coating in many fields of electromagnetic wave absorption.

2 Experimental part

2.1 Preparation

The Ni plating was repeated hydrothermal mixture to increase the Ni content after the Ni-initially Ni-coated Al_2O_3 particles were heat-treated at 1200℃ in N_2. The Al_2O_3 particles having an average particle size of 20 nm are also coated with Co particles. The Ni and Co coated Al_2O_3 moiety was heat treated at 1200℃ in N_2 and coated with Ni and Co particles. A hydrothermal mixture of Al_2O_3 and Ni and Co particles was prepared using a rotary inclined cylinder including a stone basin to compare the micro-structure of the sintered body with the micro-structure of the Ni and Co coated Al_2O_3 particles.

2.2 Characterization

The nano-Al_20_3 powders used in this experiment was about 20 nm with 99.9 % purity, other reagents also had 99.9 % purity. Take 10 ml Ni_2SO_4 (0.2 mol /l) and 2 g alumina powders hydrothermal reaction for 12 hours, and take 10 ml $CoCl_2$ (0.2 mol / l) and 2 g alumina powders hydrothermal reaction for 12 hours as a contrast group. The appearance of powders was studied by field emission scanning electron microscopy (FESEM); X-ray diffraction (XRD) pattern was used to analyze the phases of the powders, and the composition was determined by energy dispersive X-ray spectroscopy (EDS) by S-570 scanning electron microscopy. DSC-Differential Scanning calorimeters and TG-Thermogravimetric Analyzers are test by SDT Q600 Thermogravimetric Analyzer (U.S.A Ta instrument co., ltd) the test was carried out in air at a heating rate of 10 K/ min. Dielectric properties was measured by HP 8722ES microwave net work analyzer in the band of 2-18 GHz [10, 11]. For measurement of the microwave properties, the samples were dispersed in paraffin homogeneously with a sample-to-paraffin volume ratio of 1:4, and then the mixture was pressed into a toroidal shape with an inner diameter of 3.04 mm and an outer diameter of 7.00 mm. The relative complex permittivity ($\varepsilon r = \varepsilon' - j\varepsilon''$) and permeability ($\mu r = \mu' - j\mu''$) were extracted from the measured scattering parameters. The reflection loss was calculated by using the complex permittivity and permeability.

3 Results and discussion

3.1 SEM and XRD analysis

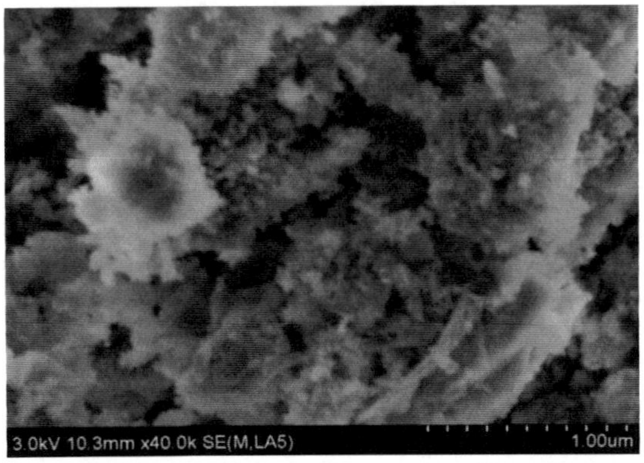

a

b

Fig. 1 a Low- and b high-magnification FESEM images of the as-prepared Ni·Al_2O_3 and Co·Al_2O_3 products. Ni:Co (2:0, 2:1)

Figure 1 shows morphology photographed by scanning electron microscope (SEM), we can see the surface of Ni / Co-alumina grains, Figure 1 (a) indicated the diameter of the powder is smaller than 30 nm. When the time of hydrothermal reaction prolonged to 12 h, the morphology of nickel layer just consisted of small irregular polyhedral particles, the nickel deposition piled up like small mound (Figure 1 (b)). All Ni-alumina grains are looks regular an fine, The average diameter of those grains probably were about 30 nm, the nickel is evenly distributed on the outer surface of the alumina.

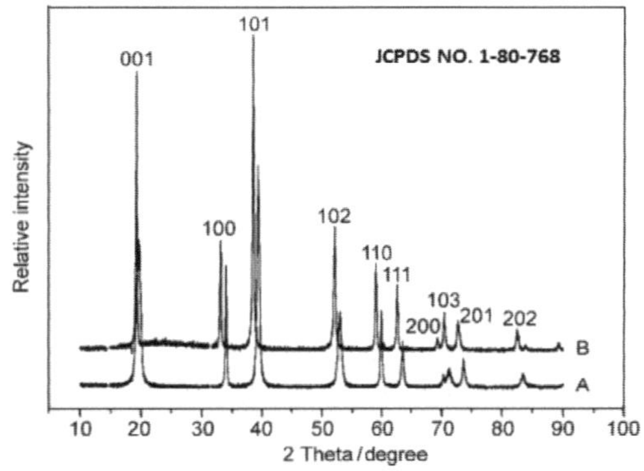

Figure 2. XRD patterns of as-prepared Ni-Al_2O_3 products a nano-particles.

X-ray diffraction (XRD) spectrum (analyzed by Jade 5.0 software) of micro-structured nickel flower-like structure film shows that crystals of sedimentary layer face-centered cubic structure (fcc) have two characteristic peaks at 44.49 ° and 51.85 ° , the lattice plane index marked (210). Since the shape of the crystal mainly depends on the relative growth rate in different directions [12], the peak is sharp, indicating that the prepared product has good crystallinity. No other impurity peaks can be detected, such as nickel oxide or nickel hydroxide, which means the purity of the synthesized nickel crystal.

3.2 Analysis of Electromagnetic Wave Absorbing and dielectric factor loss

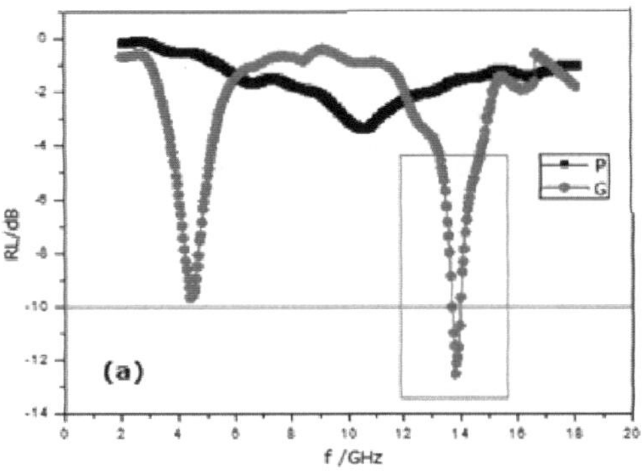

 https://doi.org/10.1051/matecconf/201818702004

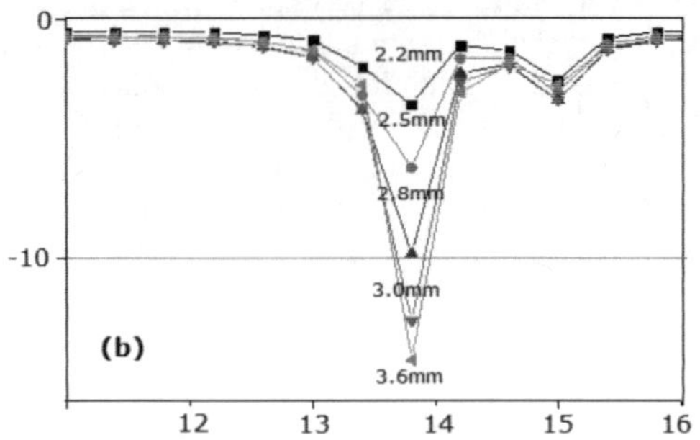

Figure 3. The electromagnetic wave absorption of 99.99% 175 mesh ultrafine aluminum powder (P) as group A and 175 mesh Ni-Al_2O_3 powder(G) as group B were measured by the reflection transmission network method at the test frequency ranged from 8 to 20 GHz. The reflection loss (RL) curves of Al_2O_3, which Sample A and Sample B have the same thicknesses. (b) is the partial magnification of (G) ranged from 11 to 16 GHz.

Finally, the microwave absorption performance should be analyzed to discuss the effect on the incorporation of nano aluminum embedded with Ni nanoparticles. The calculated reflection loss (RL) performance from the electromagnetic parameters (permittivity and permea-bility) of the as-prepared samples is shown in Figure 3. For this sample, the RL maximum is -13 dB at 13.8 GHz when thickness is 3.6 mm with an effective absorption bandwidth (RL≤10 dB) of 1.5 GHz. Additionally, the sample exhibits more excellent absorption performance than ultrafine aluminum powder in GHz frequency band.

However, the maximum RL can reach -13 dB at 13.8 GHz when thickness is 3.6 mm, which limits its application in lightweight microwave absorbing materials. In addition, when the thickness is 2.8 mm, the absorption bandwidth (RL≤-10 dB) ranges from 13.32 to 13.84 GHz. The increasing temperature can cause the skeleton collapse of frame work without the organic linkers, resulting in the attenuation of microwave absorption. Overall, among composite absorbers, Ni based composites have exhibited efficient microwave absorption [13]. It is indicated that the as-prepared materials can obtain good microwave absorption performance with thinner thickness, which corresponds to a promising potential for efficient and lightweight microwave absorbing materials.

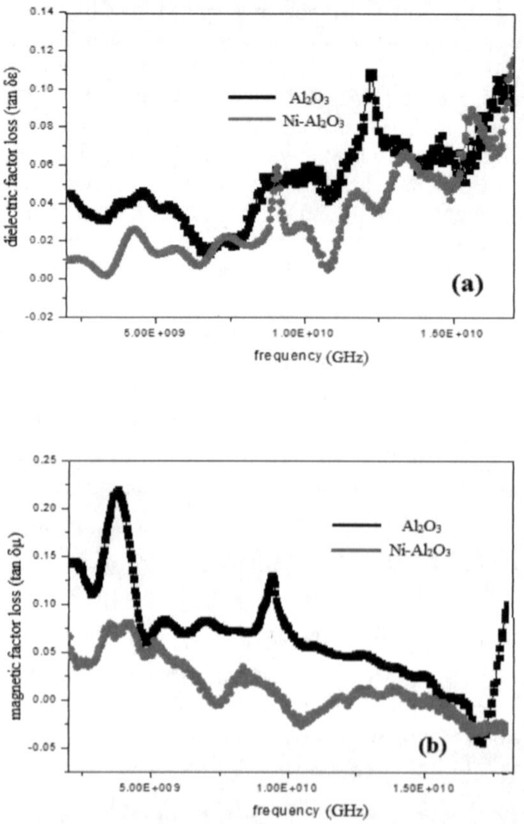

Figure 4. The dielectric factor loss (tan δε = ε''/ε') (a) and the magnetic factor loss (tan δμ = μ''/μ') (b) of Ni-Al2O3 nano-particles and solid paraffin. Attenuation constant of Al_2O_3–paraffin composites versus frequency.

The complex permittivity and complex permeability of Ni-Al_2O_3 nano-particles and solid paraffin prepared by mass ratio of 1: 1 were measured [14]. Solid paraffin is an insulating, non-magnetic material, it is a transparent material for electromagnetic waves. Solid paraffin in the complex permittivity of the imaginary part and the complex permeability of the imaginary parts are 0, solid paraffin in the composite material only play a role in the matrix and binder [15]. It can be concluded that the electrical loss and magnetic loss of the composites are due to the interaction between the Ni nanowires and the nanoparticles.

the attenuation constant can be calculated using the following equation:

$$\alpha=\left(\sqrt{2}\pi f/c\right)\times\sqrt{\left(\mu''\varepsilon''-\mu'\varepsilon'\right)+\sqrt{\left(\mu''\varepsilon''-\mu'\varepsilon'\right)^2+\left(\mu''\varepsilon''+\mu'\varepsilon'\right)^2}} \quad (1)$$

The relationship between the real and imaginary parts of the complex permittivity of Ni-Al_2O_3 nanopowder composites in the frequency range of 2-18 GHz is shown in Figure 4. However, visible from Figure 4, the real part of the dielectric constant of Nickel nanowire paraffin composites is much larger than that of Ni-Al_2O_3 / nano powder / solid paraffin composites. From the -Al_2O_3 nanoparticles / paraffin composite complex dielectric factor loss, it can be seen from dielectric factor loss part at 8 GHz and 16 GHz on the frequency range of 2-12

GHz, respectively, there were several peaks. The cause of the these peaks may be due to the quantum size effect of nano-particles, who makes the electronic energy level separation, In addition, they may due to electromagnetic scattering from the thermal motion of the electric field, impurities, defects as well as electronic and electrons interaction, all may be the cause of appearing multiple absorption peaks in the microwave band.

3.3 Thermogravimetric analysis

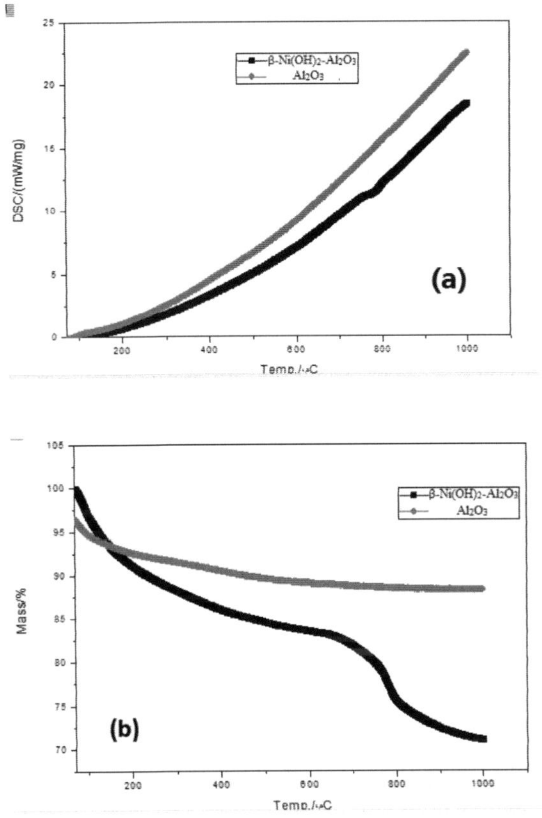

Figure 5. The DSC (a) and TGA (b) curve are related to that conversion process of β-$Ni(OH)_2$ nanocrystals to Ni-Al_2O_3 nanoparticles.

The thermal decomposition process DSC and TGA curve of β-$Ni(OH)_2$-Al_2O_3 was as seen in Figure 5. For Al_2O_3 powder, that onset of mass los in the TG trace occurs at 1000°C, which stand for the peak position in the DSC trace, as seen in Figure 5(a). The total mass loss is slightly less than the theoretical value of 7.4 %, indicating that the thermal decomposition of Al_2O_3 is incomplete. The value of the nanoparticles was 30.2 % corresponding to the formation of pure β-$Ni(OH)_2$-Al_2O_3, but the amount of residual β-$Ni(OH)_2$-Al_2O_3 was lower than the XRD detection limit for Ni-Al_2O_3, as seen in Figure 5(b). The total water loss of Al_2O_3 powder is lower than the total water loss of the Ni-Al_2O_3 nanoparticles, it should be noted that, the specific surface area is become larger because of the composition of smaller fine holes.

3 Conclusions

The structure and morphology of Ni-Al_2O_3 powder were analyzed. the results show that the Ni-Al_2O_3 nano-powder prepared by hydrothermal method has a face-centered cubic single crystal structure and the grain size is about 30 nm. The real part of that dielectric constant of nickel nanowire / solid paraffin composite is much larger than the real part of the dielectric constant of Ni-Al_2O_3 nano-powder / solid paraffin composite, and the imaginary part ε " of the dielectric constant peaks in the range of 3 - 6 GHz and 15 - 17 GHz. The real part of the permeability of the two composites decreases first and then increases with the increase of frequency, and the minimum value is obtained in the range of 8 - 14 GHz. for Ni-Al_2O_3, the electromagnetic wave absorptivity appears two peaks at 5 - 6g Hz, 17 - 18 GHz, 4 - 5 GHz and 16 - 17 GHz, respectively.

This work was supported by:

[A Project Funded by the Priority Academic Program Development of Jiangsu Higher Education Institutions （PAPD）] number:KYLX15_0310.

The funded Joint Laboratory of southern United Technology - the new electronic materials

Funding of Jiangsu Innovation Program for Graduate Education

References

1. Maoqiong, L. I. Materials Review **16**.9(2002):15-17.
2. Dishovsky, N., and M. Grigorova. Materials Research Bulletin 35, **3**(2000):403-409.
3. Duan, Yuping et. al. Journal of Physics D:applied Physics 41, **12**(2008):1854-1862.
4. Bregar, Vladimir B. Magnetics IEEE Transactions on 36, **7**(2005):1679-1684.
5. Tang, Xin, et al. Materials Science & Engineering A 445, **6**(2007):135-140.
6. Chung, D. D. L. Journal of Materials Engineering & Performance 9, **3**(2000):350-354.
7. Kaynak, Akif. Materials Research Bulletin 31, **7**(1996):845-860.
8. Giannakopoulou, T., A. Oikonomou, and G. Kordas. Journal of Magnetism & Magnetic Materials 271, **2**(2004):224-229.
9. Yan, L., et al. Nanotechnology 21, **9**(2010):095708.
10. Nicolson A M and Ross G. F. IEEE Trans. Instrum. Meas **19**(1970), 377.
11. Weir W. B. Proc. IEEE 62, **33**(1974).
12. HerbertGiesche. Journal of Dispersion Science & Technology 19, **2-3**(1998):249-265.
13. ChaoSun, et al. Journal of Materials Research 17, **5**(2002):1232-1236.
14. George W. Wagner,,†, L. R. P. †. And, and §. Shekar Munavalli‡. J.phys.chem.c 112, **26**(2007):9962-9962.
15. Wang, J., et al. Physical Chemistry Chemical Physics 17, **5**(2015):3802-3812.

Study of the Mix Cement Properties of Mortar Cement Used in Masonry and Plaster from the Waste Biscuit Firing of Ceramic

Wiwat Klangvijit [1] and Krissadang Sookramoon [2]

[1] *Department of Electronic Communication and Computer Technology, Faculty of Industrial Technology Vallaya Alongkorn Rajabhat University 1 Moo 20 Tambol Klongnueng Amphoe Klongluang PathumThani, Thailand*
[2] *Department of Mechanical Engineering Technology, Faculty of Industrial Technology Vallaya Alongkorn Rajabhat University 1 Moo 20 Tambol Klongnueng Amphoe Klongluang PathumThani, Thailand*

Abstract. The purpose of this study is to study the characteristics of cement mixes and to find out the right qualities according to Mortar Cement in terms of flow formation period and compressive strength. This work was tested for the properties of cement Mortar at Research and Development Department, Department of Concrete Testing, Royal Irrigation Thailand in ASATM C 91-97C, and ASTM C 807-89 standard. The results of the study indicated that the formula of Mortar Cement A1 flow characteristics 112, the formation time was 45 minutes and the compression strength was on the average of 39.35 MPa. Mortar Cement A2 flow characteristics 112 formation time was 45 minutes, compression strength was on the average of 25.35 MPa. Mortar Cement A3 properties of flow 118, formation time was 60 minutes, Compression strength was on the average of 17.73 MPa. Mortar Cement A4 features of flow 113, formation time was 55 minutes, compression strength was on the average of 15.33 MPa. Mortar cement A5 features of the flow 114, formation time was 55 minutes compression strength was on the average of 15.30 MPa. From the experimental result, we found that the best formulation is Mortar Cement A1, which has the highest characteristics in every property. From this study Molar cement, a material in the pozzolanic group, it is suitable for production and development in repairing old buildings or historic site. Mortar Concrete has less expansion, but high density of water slow rate ofdevelopment due to slow reaction with water. However, the compressive strength in the back is the same, the old work with the repair parts do not crack because of the surface tension, the material affected by rapid dehydration.

1 Introduction

Mortar or "ready-mixed cement" is a mixture of cement, sand and special additives in the appropriate ratio from the factory. Just rinse the bag and add water to the proportion that you can use it. This cement has a variety of formulas. The multi purpose use such as; plastering, multi-purpose repair, floor leveling, waterproofing and bonding of concrete walls with a worthy adhesion to concrete walls, which was not slipping. The application of this material is both inside and outside wall construction, which reduce cracking in smoother walls for concrete plastering. These materials, such as; plaster surface, surface finish, cast concrete, gypsum board, fiber cement board is widely used in Thailand. Removes surface imperfections, such as; air bubbles, small cracks, cracks, sagging marks, sand marks on the plaster surface of the walls and ceilings, both inside and outside are well known as well. Benefits of using mortar are working faster; reduce the work process, instant water dispenser, and consistency weight all bags. Save labor to mix mortar, conveying materials and eliminate residues. Reduce problems to repair. In the process of producing ceramics, burning bricks, clay pot plant pottery dishes are also called low-burning products. We find that there are about 2 tons of waste in the process of production per month. Most of them will be ground into the powder and return to the low pottery to reduce the damage of the workpiece. The researchers are interested in developing, processing to add value and create new products. The researcher is interested to bring the waste from the low-pottery pottery to develop mortar. Because the chemical structure consists of aluminum oxide (Al_2O_3), ionic oxide (Fe_2O_3), silicon dioxide (SiO_2) and lime ($CaCO_3$), we will obtain a fine mortar with fine sand when water is added to the optimum ratio. There are many researches on this subject. Burachat Chatveera and Permphon Srinourn [1] studied the characteristics of Portland Cement Type 4 Mix with rice husk ash and limestone powder. The properties of Portland cement, mix with ground ash and limestone powder to find out the physical and the chemical properties that are water requirements. In the study of compressive strength and shrinkage of concrete, the solutions used in this study were sodium sulfate (Na_2SO_4) and magnesium sulfate ($MgSO_4$). The concentrations of these two solutions were 5% by weight, in accordance with ASTM C1012 standard, rice husk ash and limestone powder with a particle size of 29, 12 and 128 micron.

Rice husk, ash, and limestone powder in Portland cement are 0%, 20% and 40% weight of cementitious materials. The ratio of water to cement material is controlled according to the water requirement. The results showed that the compressive strength of mortar was at 110 ± 5%. Decrease in percentage of rice husk ash and limestone powder. In Portland cement at the age of 180 days, the compressive strength of all concrete was lower than the control mortar (mixture of rice husk ash and limestone powder). The compressive strength is in the highest value in concrete control. As a result of the resistance to sulfate, the substitution of rice husk, ash, and limestone powder in Portland cement are in a higher percentage of concrete loss and shrinkage. Baboo Rai, Sanjay Kumar, and Kumar, Satish [2] evaluated the compressive strength and shear strength of 1: 3 sand mortar. The natural sand was replaced with 20%, 50% and 100% by weight of rock dust, by partial displacement (15%, 20% 25% and 30%) of low calcium fly ash. The compression time and shear strength were determined within 3, 7, 28 and 50 days. The results showed that the use of fine powdered rock and fly ash was the most effective. Due to its ability to efficiently fill the microorganism and the pozzolanic reaction. G J Prasanna Venkatesh, S, S Vivek and G Dhinakaran [3] carried out the structure, which was used as a pillar and then plastered increase the adhesion of brick, sand, cement, adhesion. In the study of cement mortar repair, structural restoration and plastering wall technology of Mortar is necessary. Cement was replaced with minerals from 5% to 20% (up to 5%), Metakaolin (MK) from 10% to 30% (up to 10%) and large scale furnace slag (GGBS) from 25% to 75% (25% increases). The ratio between cement and fine sand is 1:2 for all cement, cement and mortar components. Incubation with a different temperature was 128 °C for 4 hours. The results showed that the compressive strength of the conventional drying method and the electric oven was higher than that of the conventional concrete. The cement substitution by SF 15%, MK 20% and GGBS 25% were stronger than those of both types. The purpose of the study was; to study the mixture rate of Portland cement, waste from the low-pottery, lime, fine sand for mortar used in plastering work and to study of mortar properties in accordance with ASATM C 91-97C, ASTM C807-89 standard for example; in terms of flow, time setting, and compressive strength.

2 Materials & method

Portland cement is a mixture of limestone, soil, dirt, gypsum, mined, and mixed with water to be mixed together to burn at 1000 degrees Celsius; it will be grinded black gray to grind. 320 metric tons are collected in silos. Portland cement is used to mix stone, sand and concrete water used to make foundation beams, buildings, and condominiums. Waste from the low pottery or biscuit waste from the low-voltage containers or the brick-burning plant. There will be waste from the combustion process of 2 tons per month; one part will be used to grind the soil to be mixed with new products to reduce the crack. But with chemical elements, there are aluminosilicates. Ferric Oxide Silicon oxide can be used as a mixture of mortar. When removing the waste material from the low-pressure container, and mixed with lime and Portland cement. Mortar is used in construction and plaster works. Limestone is Lime is derived from the explosion of Limestone Mountains with lime. Take the limestone to mill and find out the sieve size and put on 20 kilograms per bag. Lime is a chemical element is calcium oxide. When mixed with the ground biscuit powder and sand, it turns to be mortar. Sand originated by the erosion of sandstone in the mountains and is carried through the canal, river canals and when deposited into the belly or led into the river, so the sand sucked up and mixed with cement.

3 Methodology

Terms of flow testing:

1. Clean the surface of the table, flow clean and dry. Then place the flow test down in the middle.
2. Load the mortar in a flow test to a height of about 25 mm and protruding the ram with 20 times throughout the cut.
3. Load the mortar in another 1 layer of about 25 mm height and mortar as well as the first layer of the mortar to put the second layer of mortar to a higher.
4. After the ram has been completed. Trowel the surface. The steel trowel is almost perpendicular to the surface of the model.
5. Lift up slowly in the vertical direction. It takes time to put the mortar down until it is lifted from the surface of the table, about 1 minute flow.
6. After the lift out of Mortar Rotate the table of flow. This will lift the spool of the flow table up to 13 mm high and let loose 25 loops in 15 seconds while rotating the swivel of the flow table. Keep tables tightly closed. Do not move because it will cause erroneous test flow.
7. Mortar was put on the plate on the top of the table, the flow will be spread like a circle. Vernier caliper was used to measure four times in the center of the mortar, at a distance of 45 angles (or measured along the trailing line on the plate of the flow table) plus the reading value. 4 sessions of Calipers The reading is the percentage of the mortar diameter increased from the original (the diameter of the flow test is 10 cm).
8. For testing using Portland cement. Record flow values between 110 (+ 5, -5) by adjusting the amount of water used to mix. [4]

Cement paste setting time testing:

1. Cement paste is prepared and make a round ball quickly. Use the glove and then throw it - come 6 times, both hands about 15 cm apart.
2. Put the cement paste into the large cone by hand. Press in and then use the hand to scrape the pest to smooth at once. Then put it on the glass. With the glass plate to make the pest to overflow smoothly with the large cone edge, then use the steel girdle to pour the cement pest into the small cone smooth to the edge of the cone. The steel

tilt is oblique and smooth. If any part of the pest cement is cracked or uneven, use a steel girdle repeatedly or gently. Do not press on the surface of this cement paste.
3. The cement paste cube band (This cement paste must be placed on the glass all the time. Move it up with this sheet of glass.) Leave it in a damp room for 30 minutes without being disturbed or inflicted.
4. Pass the cone with the glass plate to the test stand. 1 m of vibration test kit Adjust the needle scale to the surface of the cement pest, then adjust the test specimen to the sink to scale (0), so that the distance between the base of the base and the base of the cone is not less than 9 mm and the next test, the distance of the standard needle is at least 6 mm away from the original image tested.
5. Release the standard needle to sink in the pestle for 30 seconds and read the sink on the scale. The intensity may be bent or damaged. Because cement pest is very weak. The needle may be impacted against the glass placed underneath the cone. Then slowly release the needle into the cement pest. This method is technically only not the correct way (the correct method is to release the needle only).
6. Consent it for 15 minutes and test it again. (In accordance with clauses 4 and 5) with time recording of the penetration of the standard needle. Do this until the sinking period. 25 m. The values are inclusive of the time from the beginning of the mixture to the duration of the 25 mm sink, called the initial formation period.
7. The formation of the end of this form would variate with the use of the needle to release the needle on the surface and then perform on the cement surface pest. The time from the start of the mix of this time is the end of the formation phase. The cement paste setting time testing is shown in Figure 1.

Figure 1. Cement paste setting time testing.

4 The compressive strength testing:

1. Cement cube mold in the size of 2x2x2 inch or 50x50x50 mm, the test form must be tightly packed. Up to two pieces can be used and one cast has a maximum of three cubes. The mold is made of solid material, smooth surface, with corner angle. Not more than 0.5 mm and not reacted with mortar. Flow Table and Flow Mold used to test the flow of mortar. Comply with ASTM Designation C-230.
2. Tamper rods must be made of non-absorbent material. It is not worn or frictionless and fragile, such as made of hard wood, then discarded. And do not absorb water by soaking in hot paraffin oil at 200 °C for 15 minutes or by hardening with a hardness of 80 ± 10 measured by a Shore A Durometer. The field must have a cross section of 13x25 mm (1/2 x 1 inch) and a length of about 120 to 150 mm and the cross section shall be perpendicular to the long axis of the rod.
3. The mixer is electrically driven. Comply with ASTM designation C-305.
4. Balance is reading 0.1 g.
5. Measuring cylinder is large enough to measure water only once. The measurement was done at 20 °C with a tolerance of not more than ± 2 m. In this test, the measuring range was 500-1000 m.
6. Trowel (flat iron) the rim of the trowel.
7. Vernier Caliper is measured in the range of 8-10 inches.
8. Hydraulic testing machine, the spiral between the front of the upper and lower keys. It must be enough space to remove the equipment needed to test the installation. And the gap for the test cube fit. Pressures must be accurate to ± 1 percent. The upper and lower keys are hard metal. The top of the keyboard has a rounded base that fits snugly against the center of the upper key. As we walked the test machine, the top was down with the specimen on top of it, although it was firmly attached to the sphere. It can be tilted in all directions according to the surface of the contact surface of the press pad, platen plate hardened using a test sample. The Rockwell hardness must not be less than HRC 60. [5] The Compressive Strength testing is shown in Figure 2.

Figure 2. The compressive strength testing.

https://doi.org/10.1051/matecconf/201818702005

Figure 3. The compressive strength testing.

5 Experimental results

Table 1 shows the properties of mortar by studying the ratio of Portland cement, low-calorie, lime, fine sand which categorized in formula A_1-A_5. Terms of flow is shown in Figure 4, The cement paste setting time is shown in Figure 5, and The Compressive strength of Portland cement is shown in Figure 6. Mortar has a standard of the flow rate of 110 (+5, -5), the formation period time is 60 minutes and a compaction capacity is 2.5 mega pascal. Formula A_1 is the best property of mortar mixing Portland cement. Formula A_2-A_5 permits the standard, excluding the formular A3 does not pass the standard because the terms of flow exceeds the regular.

Table 1. The properties of mortar by studying the ratio of Portland cement, low-calorie, lime, and fine sand which are categorized in Formula A_1-A_5.

Fomula	Portland cement %	Waste from the pottery %	Limestone %	Sand %	Terms of flow	Setting time minute	Mean Compresstive Strength Mpa
A_1	12.14	15.17	12.69	60	112	45	39.35
A_2	12.14	12.69	15.17	60	112	45	25.35
A_3	13.25	13.50	13.25	60	118	60	17.73
A_4	13.25	13.25	13.50	60	113	55	15.33
A_5	10.90	18.21	10.90	60	114	55	15.30

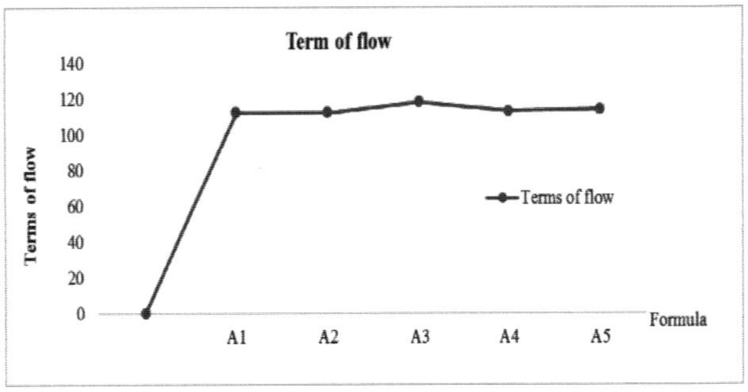

Figure 4. Term of flow of each formula.

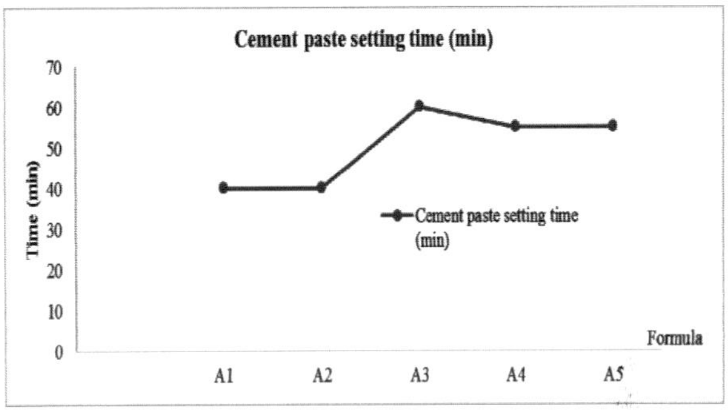

Figure 5. The cement paste setting time of each formula.

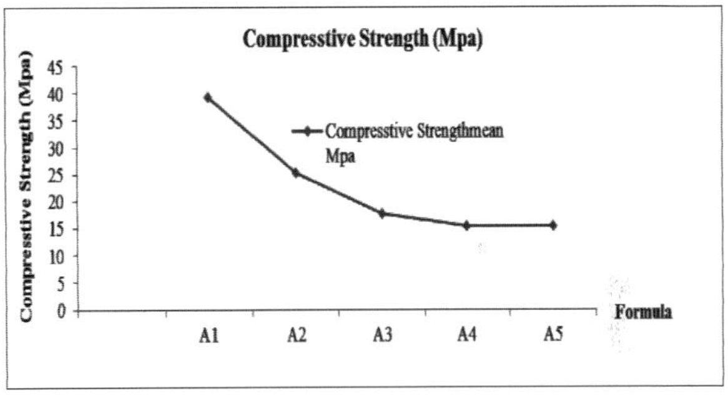

Figure 6. The Compressive strength of Portland cement of each formula.

6 Summary

The properties of mortar mixing Portland cement from the study(low-calorie, lime, fine sand) indicated that the best mixing ratio of Portland cement was 12.14%. Waste from the low pottery was 15.17% Limestone was 12.69%, and sand was 60%. The high quality of the experiment was formula A_1 in terms of flow 112 and the cement paste setting time was 45 minutes and the mean compressive strength was 39.35 Mpa. Cement mixes for material Pozzolan are suitable to be used in the repair and construction of old buildings. Concrete has a less expansion high density of water. Slow rate of development due to slow reaction with water. However, the compressive strength is the same level as the old value, the old work with the repair parts do not crack because of the surface tension, the physical affected by rapid dehydration.

References

1. Chatveera, B. and Srinourn, P. A Study of Properties of Mortar, Portland Cement Type V Mixed with Ground Rice Husk Ash and Limestone Powder. Journal of Research and Development King Mongkut's University of Technology Thonburi. Year 2 April - June (2012)
2. Baboo Rai,Sanjay Kumar,and Kumar Satish. Effect of Fly Ash on Mortar Mixes with Quarry Dust as Fine Aggregate. Hindawi Publishing Corporation Advances in Materials Science and Engineering, Volume Article ID 626425,7 pages (2014)
3. G. J Prasanna Venkatesh, S S Vivek and G Dhinakaran. Study on compressive strength of self compacting mortar cubes under normal & electric oven curing methods. ICCIEE 2017 IOP Conf. Series: Earth and Environmental Science80 012002 (2017)
4. SIRISAWAT I, SAENGSOY W, BAINGAM L, KRAMMART P. Durability and testing of Mortar with underground fly ash and limestone cements in sulfate solutions. Journal of Construction and Building Materials, Vol.64 (2014)
5. ZHANG Ming-hua, JIANG Ming-qiang, CHEN Jian-kang. Variation of flexural strength of cement mortar attacked by sulfate ions. Journal of Engineering Fracture Mechanics, Vol.75 (17) (2008), p. 4948–4957.

https://doi.org/10.1051/matecconf/201818703001

Strategies for Promoting Steelmaking Slag Resourcezation in Taiwan

Ching-Jung Chang

Department of Construction Management, Chung Hua University, Hsinchu 30012, Taiwan

Abstract. Because of the world ever-growing pursuit of economic growth, natural resources have been overly exploited and squandered. Such actions have created a negative effect on the environment and led to resource depletion. These problems subsequently garnered the government attention for waste reuse, in which they began to promote sustainable circulation and reuse of resources to lower natural resource consumptions. Accordingly, the concept of the "3 R's," namely, reduce, recycle, and reuse was introduced. However, the said concept merely adopts a unidirectional, cradle-to-grave-based treatment and fails to consider the effects of waste on the economy and society. Taiwan produces approximately 20 million tonne s of waste annually, roughly 40% of which are slags. The illegal dumping and burying of slags have negatively influenced Taiwan's environment, economy, and society. Thus, the present study revolutionized the 3R concept by introducing the cradle to cradle concept. Producers are recommended to consider product features and slag usage options during the steelmaking process. Transparent productions, effective management, scientific treatments, and systematic inspections are employed to bring slags into industrial circulations, facilitating the recycling and reuse of resources.

1. Introduction

The steel industry is the mother of all industries and the foundation of national development. However, the steelmaking process results in a significant amount of waste, known as slag waste. In 2014, steelmaking plants around the world created approximately 160 million tonnes of slags [1], in which Taiwan accounted for approximately 7.7 million tonnes of such slags annually. To achieve slag "resourcezation" (i.e., make slags into resources), one must first understand slag properties and determine what they are to be used before implementing the corresponding reuse technologies. Such endeavors will facilitate the goals of waste resourcezation, create market-based resource circulations, reduce the negative impact of slags on the ecosystem, and elevate industry operation efficiencies.

Because different raw materials and steelmaking methods are used during the different stages of the steelmaking process, they create different slag compositions and properties. This leads to significantly different resource reuse options. Concerning slags, they are divided into blast furnace (BF) slags, basic oxygen furnace (BOF) slags, desulfurization slags, electric arc furnace (EAF) oxidizing slags, and EAF reductive slags. BF, BOF, and desulfurization slags are byproducts of the continuous steelmaking process, whereas oxidizing and reductive slags are those of the EAF steelmaking process.

When smelting is performed in a blast furnace (in an integrated steeling mills) during the steelmaking process, materials such as limestones are added as fusing agents to allow iron ore and coke impurities to synthesize, forming slags. Because slags have a relatively smaller specific gravity, they can be separated from molten iron. When slags created from the blast furnace process are cooled, they produce a solid matter called BF slags. By cooling high-temperature BF slags in liquid state using different cooling methods, air-cooled BF slag and ground granulated BF slags can be produced. Because molten iron created from BF-based smelting contains sulfur, it is transferred to torpedo to be desulfurized by desulfurization agents. The resulting product is naturally cooled to form a finely grained solid called desulfurization slags. By contrast, when smelting is performed in a converter during the steelmaking process, it creates Basic Oxygen Furnace slag (BOF slags). Because the said process involves making iron ores into iron first before making them into steel, it is called continuous steelmaking.

For electric arc furnace steelmaking, the primary raw materials used are recycled scrap steel and iron. Regarding slags created from the EAF steelmaking process, they comprise EAF oxidizing slags (created prior to the steelmaking process) and EAF reductive slags (created after the steelmaking process). Because the sources of recycled scrap steel and iron differ, and that minerals added during the steelmaking process vary, each batch of slags created exhibits significantly different chemical composition and quality.

Every year, steelmaking plants discharge million tonnes of toxic substances into the natural environment, consume and waste countless resources, produce innumerable waste, destroy the global cultural diversity, and jeopardize biodiversity [3]. To make up for these mistakes, government agencies, researches and environmentalists proposed the concept of the 3 R's. However, such a concept did not eliminate the discharge of toxic substances or the consumption of resources and merely extended product life cycles by two or three-fold. The end result remained the same: cradle to grave.

Thus, Professor Braungart put forward a proactive solution, positing that reduction in damages created will not make the world a better place and that the concept of waste does not exist in nature. In essence, all substances should be viewed as nutrients that can be returned to nature. To achieve this ideal, he asked designers to consider the methods to transition resources used in product production into the beginning of the next cycle, and maintained that designers should do this as early as the product design stage. He termed this idea "cradle to cradle".

The paper examined the promotional strategies employed to resourcize various types of steelmaking slags, integrated relevant treatment technologies, and adopted the cradle to cradle concept to achieve the goals of slag resourcezation and improve slag recycling. This study put forward what the various parties of the steelmaking process could do to manage slags and facilitate the use of slags in the industrial "metabolic cycle." The study results may serve as a reference to producers, consumers, and the government when implementing environmental protection strategies.

2. Slag classifications and properties

Waste is defined generally as an unusable substance or one of no value [4]. As environmental protection awareness becomes increasingly important and natural resources become increasingly scarce, identifying approaches to utilize reprocessing procedure and proper planning to properly reuse waste, convert identified waste into recycled resources. Transform waste into industrial materials are crucial to effectively lowering the misuse or disposal of waste and thereby diminishing its negative impact on the environment. Because of the numerous types of waste that exist and that their disposal or retreatments differ. For waste reuse to be feasible, the following conditions must be met: the waste materials must be able to stably serve as alternatives of materials demanded by the market, and the demand for waste materials must be greater than their supply. When these two conditions are met can an incentive for using waste be created and waste quantity and storage be reduced.

Slags contain silicates and ferrite compounds made of SiO_2, Al_2O_3, CaO, MgO, and Fe_2O_3. Because slags have a chemical composition similar to that of cement, they should primarily be used as raw cement materials and/or cement replacements in cement-related applications and as backfill and/or aggregate replacements in civil engineering-related applications. Slags are divided into six categories according to the steelmaking process employed and the resulting slags produced. The slag production processes and properties are explained as follows:

(A) Slags produced from continuous steelmaking plants

A.1. Blast furnace slags

BF slag is formed in a blast furnace with molten iron from iron ore, once removed from the furnace, are subject to further processing to form a crystalline aggregate or ground blast furnace slag.

A.1.1. Air-cooled BF slags

Air-cooled BF slags, cooled naturally by air and formed form a crystalline aggregate, particle size of up to several meters [5]. Because high-temperature liquids and flowing, they envelop air during the cooling process. Once cooled, they show a porous and uneven surface. They do not contain harmful substances such as heavy metals and dioxin.

A.1.2. Ground granulated BF slags

Ground granulated BF slags are created via a high-pressure water quenching process, account for approximately 95% of all BF slags, and display hydrate cementation characteristics. The said slags are ground into ground granulated furnace powder, which is mixed with Portland cement to form Portland BF cement.

A.2. Basic oxygen furnace slags

Molten iron created in the BF is transferred to converter, in which raw materials such as limestones are added as fusing agents to remove molten iron impurities during the molten steel blowing process, forming molten steel. The slags created from the aforementioned process are BOF slags. BOF slags consist primarily of chemical compounds CaO, Fe_2O_3, SiO_2, and MgO, and chemical stability tests showed that they are nontoxic and do not contain dioxin.

A.3. Desulfurization slags

Molten iron created from smelting in a BF contains sulfur content. The molten iron is subsequently transferred to the torpedo to be desulfurized by desulfurization agents. Once discharged from the torpedo, the slags undergo natural cooling and form finely grained solid matters called "desulfurization slags."

(B) EAF slags

The primary raw materials used in steelmaking are scrap steel and iron. Following a smelting reaction, most nonferrous metals and added auxiliary raw materials create EAF slags. Concerning EAFs, they are primarily made of three artificial graphite electrodes, a furnace ceiling, and a furnace body. When charged with electric current, the graphite electrodes generate electric arcs that measure at least 1,600°C, melting scrap steel and iron and completing the steelmaking process. Steelmaking operations are divided in batches and each batch normally runs 1.5–3 hr. Regarding the smelting process, it is divided into three stages, which are the melting stage, the oxidation stage, and the reduction stage. The slags created during the aforementioned stages are classified as oxidizing and reductive slags, as explained below:

B.1. Oxidizing slags

Scrap steel melted by electric arcs becomes molten steel. The recycling of scrap steel is a rather complex process, in which molten steel impurities have to be oxidized to turn into liquid-state or gas-state oxides and high-pressure oxygen has to be added to accelerate the oxidation effect. The goal is to expose the molten steel impurities to oxygen to form oxides. Slags created during this process are called oxidizing slags, which have a rough shape and resemble natural igneous rocks. Oxidizing slags are primarily made of SiO_2, Al_2O_3, Fe_2O_3, CaO, MgO, and S and have a metal content of approximately 50%, in which Fe_2O_3 and Al_2O_3 account for the highest ratio.

B.2. Reductive slags

To prevent molten steel from containing too much oxides and affecting steel quality, reducing agents such as siliceous ores, ferromanganese, cokes, and quicklime are added to control the motel steel quality. The main objective is to induce a reaction with oxides to facilitate deoxygenation and desulfurization. Slags produced from this stage are called reductive slags. Reductive slags have a small powder-like appearance and are grayish white in color. Reductive slags are mostly made of CaO (approximately 48%), SiO_2 (approximately 22%–29%), Al_2O_3 (approximately 14%–18 %), and MgO.

3. Cradle to cradle

The companies always set economic growth as the ultimate goal of their countries. Accordingly, the focus of product designs and manufacturing has been to identify ways to deliver products to consumers quickly and inexpensively. Issues such as product life cycles and the negative impact of products on the environment were ignored [3]. Natural resources were used for unidirectional, cradle-to-grave-based product development, in which the processes involved processing resources, manufacturing products, using resources, disposing resources, and polluting the environment. In response to the said problems, the 3 R's (i.e., reduce, recycle, and reuse) were introduced. However, such methods were overly passive and did not solve the problem at the core. Strategies that emphasized reduction and recycling merely extended product life cycles and reduced usage. By contrast, Professor Braungart put forward the concept of "cradle to cradle" in 2002, contending that the idea of diminishing damages made to the environment is overly naive, and that such plans to save the Earth's natural environment would fail. He proposed a proactive solution, which focused not on repairing the damages created by industrial productions, but rather on learning how to recycle sustainably.

In Professor Braungart cradle to cradle analysis, the Earth is presumed to have two distinct metabolisms [3,6,7], namely, a biological metabolism and an industrial metabolism, and the following three basic principles are in effect: (1) view waste as food (every substance is the nutrient for another substance); (2) use existing solar power; and (3) respect diversity (i.e., biodiversity, concept diversity, and cultural diversity).

3.1. Two distinct metabolisms

3.1.1. Biological metabolism

The concept of waste does not exist in nature. For example, flowers and fruits of cherry trees that fall to the ground return to the soil and become nutrients in the ecosystem, in which nothing is wasted. Similarly, plants that die at the end of their life cycle decompose in the soil and serve as nutrients for the ecosystem to circulate.

3.1.2. Industrial metabolism/circulation

Designers who think about how different raw materials can be introduced to different circulations during the product development stages will allow the materials to retain their original properties and even achieve "upcycling." Because industrial products normally cannot decompose naturally, they should be used in industrial circulations, in which they are recycled and remade into valuable industrial nutrients for subsequent use.

3.2. *Three* basic principles

1. Waste is food: Every substance in nature is the nutrient for another substance. Similarly, industrial products should be upcycled.

2. Use existing solar power: Reduction in resource usage does not prevent the creation of waste or the consumption of limited energy-generation resources such as coals, petroleum, and nuclear power-related resources. By converting natural energy resources such as the Sun and winds into power, nature, the Earth, and the Earth's resources can be protected.

3. Respect diversity: Respect for biodiversity, material diversity, and cultural diversity should be practiced prior to designing and producing products. Sustainable development should be in line with local conditions, in which local materials, culture, and ecology are accounted for during product developments. To truly achieve sustainability, product manufacturers must not be short-sighted by favoring the use of cheaper energy resources and not taking into account future treatment costs.

Ecological efficiency is only used in the context of

product life cycles, in which the effects of the various stages of products on the environment are assessed and the emphasis on pollution reduction is made. Professor Braungart argued that merely improving ecological efficiency is insufficient [8] as this only makes the matter "less bad" but not better. By contrast, cradle to cradle involves putting eco-effectiveness into practice. Such a practice involves implementing positive, step-by-step procedures rather than adopting unrealistic methods aimed to achieve perfection in a single attempt.

4. Analysis of slag strategies

In Taiwan, slag users having a lack of expertise and intransparent slag usage, producers implementing loose management, managers lacking complete verification systems, related units treating slags as waste and burying them or throwing them away. Companies misusing slags have resulted in the general public and engineering units misunderstanding of slags, making the promotion of slag resourcezation difficult. Therefore, the paper implemented the circulation-oriented concept of cradle to cradle, in which slags were treated as industry nutrients and the chemical and engineering properties of the different types of slags were analyzed to determine their respective resourcezation triage strategies. The goal was to allow the slags to become a part of the industrial metabolism. By treating toxic substances and making them into industrial nutrients, not only are the use of resources reduced, but the toxic substances are upcycled, enabling slags to be effectively reused and remain in closed industrial circulations.

With respect to sustainable slag developments, three dimensions are considered: (1) Economic dimension (i.e., balance): Slags are upcycled to allow them to be sold at a low price and used as replacements of natural resources; (2) Social dimension (i.e., sustainability): Users are provided with safe slags that they can use at ease; and (3) Environmental dimension (i.e., balance): Slags are used as replacements of engineering materials without polluting the environment. To achieve the aforementioned objectives, properties of the different slags and their usage potentials have to be identified, the annual output of various slags as well as users' demands for such slags have to be obtained, the market in which the slags can be used as alternative materials demanded by the market (in which the demand for the slags have to be greater than their supply) have to be determined, the amount of slags eliminated and where they are disposed have to be made transparent, and government policies have to be complied and promoted, in which producers are asked to reach a consensus.

To successfully promote slag reuse, diversified strategies will be implemented. For example, civil engineering and/or cement-related producers will be recruited to convert slags into those featuring specifications demanded by users. Next, in accordance with government laws and regulations, complete and transparent product channels will be introduced to ensure that slags are treated by designated units and that their usage and disposal locations are controlled.

4.1. Properties of the different slags

4.1.1. Blast furnace slag

A. Air-cooled BF slags: Because Air-cooled BF slags which have a large particle size with porous and uneven surface do not contain harmful substances such as heavy metals and dioxin, they could be used as engineering materials after being crushed and sieved. They may be used as roadbeds, concrete aggregates, general strength structural concrete, nonstructural concrete (e.g., ditch covers, curbs, high-pressure bricks, and water-permeable bricks), and controlled low strength concrete (CLSM).

B. Ground granulated BF slags: Ground granulated BF slags are created from a high-pressure water quenching process. With a main chemical composition of Ca0, SiO_2, and Al_2O_3, they are similar to Portland cement and display hydrate cementation characteristics. When ground into powder, ground granulated BF slags have the potential to exhibit cement-like abilities. By mixing furnace powder with Portland cement, it creates Portland BF cement, which is the most mature and profitable of all slags. Concerning fine particle BF powder, it can be used as a fill material and blended with concrete to reduce cement slurry pores and increase concrete density, which subsequently enhance concrete durability and quality.

4.1.2. Basic oxygen furnace slags

BOF slags display the following physical properties: (1) durability; (2) heat-holding ability; (3) friction properties; (4) anti-flaking ability; (5) stability; and (6) anti-rutting ability [9]. Regarding the recycling and resourcezation of BOF slags, BOF commonly used as raw cement materials, backfill materials, and soil modifiers as well as materials in road surface and base layers, soil improvements agent and water conservancy projects. Japan used BOF slags to reduce, prevent, and control diseases in soil used to grow vegetables, in which it obtained significant and positive results. Germany used fertilizers that contained BOF slags to grow grass and cereals, in which the test results showed that the yield increased by 10%–20% compared with those that used general phosphate fertilizers. In addition, contaminations such as the leaching of heavy metals such as cadmium(Cd) were not observed [5, 10].

BOF slags undergo processing procedure such as crushing, magnetic separation, and sieving to become outstanding recycled engineering material-class aggregates. However, the said slags contain free CaO, which are prone to absorbing water or moisture, which induces $CaO+H_2O \rightarrow Ca(OH)_2$ reactions. When such reactions occur, BOF slags expand, dissolve, and turn into powder. Because materials used in roadbeds and road base layers must have an expansion rate of less than 1.5%, for BOF slags to meet this requirement, the splash cooling method may be replaced with the ballast water pocket splitting method during the slag cooling process. The new method allows hot slags to soak in water and stabilize, significantly lowering the expansion rate.

4.1.3. Desulfurization slags

Molten iron created in a BF contains sulfur. To desulfurize molten iron, desulfurization agents are added. This process creates fine particle slags called desulfurization slags. Concerning desulfurization slag resourcezation, they are mostly used as ironmaking materials or low strength concrete. Desulfurization slags that are purified after magnetic separation have a pH value of 10–12 and may be favorably used as raw cement materials, low strength concrete, cement product fill materials, brick-making materials, coarse aggregate replacements, earth backfill, soil modifier raw materials, and limestone replacements in fired cement clinkers [11,12].

4.1.4. EAF oxidizing slags

Oxidizing slags are mostly dark brown in color. Concerning their physical properties, they are durable and wear-resistant and display a low water absorption rate (1.5%–6.8%), a high specific gravity (2.5–3.7), high particle strength, low abrasion consumption, and high water permeability. In addition, they do not release toxic substances into the environment, are stable in size, and are suitable as pavement materials and coarse aggregate replacements.

Because pavement materials and coarse aggregate replacements normally contain a high iron content, they lead to surface corrosion problems. To solve such problems, magnetic separators may be used to remove the iron content and preserve the overall appearance of the pavement. Oxidizing slags demonstrate favorable durability and anti-rutting ability and are now widely used in the asphalt pavement of major traffic intersections in Taiwan and Singapore [2, 5].

4.1.5. EAF reductive slags

Reductive slags have a small powder-like appearance and are grayish white in color, because a portion of their f-CaO and MgO will become unstable and expand during reuse. Stabilization treatments must be performed prior to their reuse to enhance their stability [13].

4.2. Industrial application of different slags

To achieve waste reuse, the waste must be able to stably serve as alternatives of materials demanded by the market, and the demand for waste materials must be greater than their supply. The use of slags as alternative materials generally takes place in civil engineering-related industries. In this study, the reuse strategies adopted for seven major engineering projects were analyzed. Materials used for such projects are listed as follows: (1) road base layer; (2) asphalt concrete pavement; (3) ready-mixed concrete aggregates; (4) controlled low strength concrete (CLSM); (5) cement products (concrete bricks, manholes, ditch covers, curbs, and concrete pipes); (6) raw cement material (alternative admixture and raw material); and (7) backfill and landfill.

The concept of cradle to cradle was implemented, in which what the slags could serve as replacements of were determined according to their properties. This allowed the slags to be recycled and reprocessed accordingly and made into valuable industrial nutrients to be used in closed industrial circulations. Table 1 shows Taiwan's annual output of the various types of slags. According to the table, the demands for the slags as alternatives of materials for the seven major engineering projects all exceed their supply, meeting the cradle-to-cradle industrial metabolism criterion, allowing raw materials to maintain their original properties when introduced to different circulations, achieving the goal of upcycling, and satisfying the principle of waste is food. However, some of the related technologies remained immature and are to be improved as new industrial technologies are made available the future.

Table 1. Taiwan's annual output of the various types of slags [2, 13, 14]

Slag type	Byproduct(s) created and the steelmaking process involved	Annual output (unit: 10,000 tonnes)	Mainly used as a replacement of	materials demanded (unit: 10,000 tonnes/year)	Usage restriction assessment
1	BF slags created from the continuous steelmaking process	400			
1.1	Air-cooled BF slags :account for approximately 5% of all BF slags	20	1.Road base layer	150[1] :*	
			2.Asphalt concrete pavement	600[2]	
			3.Ready-mixed concrete aggregate	9250[3]	
			4.Controlled low strength concrete	250[4]	
			5.Cement products	200[5]	
1.2	Ground granulated BF slags: account for approximately 95% of all BF slags	380	1.Raw cement materials	1200[6]	
2	BOF slags created from the continuous steelmaking process	150	1.Road base layer	150[1]	Stabilization treatment required
			2.Asphalt concrete pavement	600[2]	Stabilization treatment required
			3.Ready-mixed concrete aggregate	9250[3]	
			4.Controlled low strength concrete	250[4]	
			5.Cement products	200[5]	
			6.Raw cement materials	1200[6]	
			7.Backfill and landfill	460[7]	
			8.Stell slag powder		
3	Desulfurization slags created from the continuous steelmaking process	40	1.Road base layer	150[1]	
			2.Asphalt concrete pavement	600[2]	
			3.Ready-mixed concrete aggregate	9250[3]	
			4.Controlled low strength concrete	250[4]	
			5.Cement products	200[5]	

			6.Raw cement materials	1200[6]	
			7.Backfill and landfill	460[7]	
4	EAF oxidizing slags created from the EF steelmaking process	140	1.Road base layer	150[1] :	
			2.Asphalt concrete pavement	600[2]	Improve corrosion by removing iron using a magnetic separator
			3.Ready-mixed concrete aggregate	9250[3]	Improve corrosion by removing iron using a magnetic separator
			4.Conrolled low strength concrete	250[4]	
			5.Cement products	200[5]	
			6.Backfill and landfill	460[7]	
5	EAF reductive slags created from the EAF steelmaking process	40	1. Cement products	200[5]	Stabilization treatment required
			2.RAW cement materials	1200[6]	Stabilization treatment required

*: $X^Y \rightarrow$ X: the demand for materials that the slags can be a replacement of; Y: the replacement path and number

5. Conclusion

1. Prior to reusing slags, their assessments must be made to identify their properties. Next, appropriate usage options must be determined accordingly. Slags created from the EAF oxidation stage and those created from the reduction stage show considerably different chemical compositions and characteristics. Because reductive slags and BOF slags contain relatively more free CaO and MgO, they create gradual expansion problems when used in backfill or asphalt pavement. Thus, prior to such usage, stabilization treatments must be performed. Concerning oxidizing slags, because they have a high iron content, which leads to surface corrosion, magnetic separators must be used to remove the iron content and improve corrosion.

2. The annual outputs of slags are relatively constant. To prevent unauthorized burying of slags, their disposal locations must be transparent and their disposal must be managed by designated units. In addition, complete technologies must be used jointly to facilitate waste resourcezation. Furthermore, standards and verification systems must be enforced according to government policies to allow the professional industry, the government, and the academic industry to promote cradle to cradle together.

3. To allow slags to be used as recycled resources, diversified strategies should be devised. Various aspects should be taken into account, such as slags displaying chemical properties similar to cement products and physical properties similar to coarse and fine aggregates used in concrete, the large demand for civil engineering materials, and the diversified applications of civil engineering materials. On the basis of the slag properties, different, most appropriate reuse options should be formulated and accurate slag yields and the total demand for slags as replacement materials should be obtained. Finally, the demand for waste materials being greater than their supply should be confirmed to ensure sustainable usage and achieve stable slag elimination.

4. The ultimate goal of slag resourcezation is to implement the cradle to cradle concept, achieve upcycling, reuse waste, maintain waste in the industrial metabolism, and realize sustainable economic, social, and environmental balance.

6. References

1. P.T. Teo, A.A. Semana, P Basub, and N.M Sharif, "Characterization of EAF Steel Slag Waste: The Potential Green Resource for Ceramic Tile Production". Procedia Chemistry (2016), P.842–846
2. Taiwan Construction Research Institute. "An

Introduction to the Properties of Steelmaking Slags and Related Management Laws, Regulations, Standards, and Guidelines." Seminar and Collected Papers on the Optimal Method for Achieving Steelmaking Slag Resourcezation and Application, Management, and Control Strategies for Concrete Engineering. (2016)

3. W. McDonough M. Braungart , *Cradle to Cradle: Remaking The Way We Make Things*. New York, North Point Press, 2002
4. S.Z. Chang, "Physical and Chemical Properties of Electric Arc Furnace Oxidizing Slags and Mortars." Master's Thesis. Institute of Construction Management, Chung Hua University; Hsinchu. (2015)
5. CHC Resources. Access date 7 Feb. 2017; available from http://www.chc.com.tw/product_2.html
6. M. Braungart, W. McDonough, *The Upcycle: Beyond Sustainability–Designing For Abundance*. New York, North Point Press, 2013
7. MBDC LLC. Cradle to Cradle CertifiedCM Product Standard Version 3, Access date 7 Feb. 2017; from http://www.c2ccertified.org/images/uploads/C2CCertified_Product_Standard_V3_121112.pdf
8. M.E. Toxopeusa, B.L.A de Koeijera, A.G.G.H. Meijb. Cradle to Cradle: Effective Vision vs. Efficient Practice? The 22th CIRP International Conference on Lifecycle Engineering, Sydney Australia.Heidelberg, Springer, 2015 Procedia CIRP 29 P.384 – 389, 2015
9. D. Huang, "The Investigation of Replacing Natural Aggregates with BOF Steel Slag which Effecting VMA Characteristics of Hot Mix Asphalt" Master's Thesis, National Central University; Zhongli, 2008.
10. X.R. Chen, and G.Z. Chang-jian, "The Effect of Basic Oxygen Furnace Slags on Environmental Compatibility." *2011 Seminar and Collected Papers on the Use of Basic Oxygen Furnace Slags in Asphalt Concrete Pavement.* (2011)
11. S. K. Kawatra, S. J. Ripke, "Pelletizing steel mill desulfurization slag", International Journal of Mineral Processing, **vol. 65**. Elsevier, Amsterdam, pp.165–172, 2002
12. W.T. Guo, and S.S. Chen, "The Feasibility of Defulsurzation Slag Used as Aggregates for Asphalt Concrete". *Journal of the Chinese Institute of Mining & Metallurgical Engineers*, **Vol. 54, No.4,** pp. 40-50, 2010.
13. S.W. Tseng. "Assessing the Use of Electric Arc Furnace Reductive Slags as Controlled Low Strength Materials and Their Stabilization Performance." Master's Thesis. National Central University, Zhongli. 2012
14. J.C. Xing "Utilization of the Electric Arc Furnace Oxidizing Slag as Microstructure Materials." Master's Thesis (2014). Institute of Materials and Mineral Resources Engineering, National Taipei University of Technology: Taipei, 2014

Syngas from Updraft Gasifier Incineration for Internal Combustion Engine Power Generation in Klongluang PathumThani Thailand

Krissadang Sookramoon

Faculty of Industrial Technology Vallaya Alongkorn Rajabhat University 1 Moo 20 TambolKlongnueng Amphoe Klongluang PathumThani, Thailand

Abstract. This paper presents the internal combustion engine power generation using syngas from the updraft biomass gasifier as a fuel. 3 types of fuel such as Golden shower tree wood chip, charcoal, and gasohol 91 were tested for the engine running. The experiment was performed on July 25-26, 2017 at Faculty of Industrial Technology Vallaya Alongkorn Rajabhat University Pathum Tani Thailand. Data on the performance of the engines fueled with producer gas and gasohol 91 is presented. The experiment was carried out by using a four-stroke 13 HP gasoline engine coupled with a generator as a load in producing electricity. The carburetor was modified for fuel gas running by loading 7 kg/batch of Golden shower chips and charcoal for syngas producing and the engine performance was measured. The results showed that, the engine power was 110.125 W, 115.425 W, and 128.038 W, while using a golden shower chip, charcoal, and gasohol 91 as the fuel, respectively. The generator efficiency is 80% therefore the generator power reduces 20%. The test indicated that golden shower chips could produce higher quality of syngas than charcoal but the engine power has less power than fueled with gasohol 91.

1 Introduction

The use of biomass energy as a renewable energy such as charcoal, rice husk or in the form of scrap, which are locally available resource. The syngas is composed of carbon monoxide (CO), hydrogen (H_2) and methane (CH_4), which can be used as fuel instead of petroleum fuel oil in the internal combustion engine. Although biomass gasification is considered as an ancient technology, the consequence of the use of this technology has increased in order to preserve fossil fuel resources and decrease greenhouse gas productions. The removing of tar from producer gas has been found to be a major problem for engine applications. It can be used as fuel instead of petroleum fuel in internal combustion engine such as gasoline engines both large and small. The performance test of a gasifier used for this small engine can be used as a machine in place to generate electricity, pumping water, etc., driving the machines in the agricultural sector.

Golden shower tree, in other word Cassia fistula is a fast-growing, medium-sized, deciduous tree which is now widely cultivated in Thailand as an ornamental tree for its attractive dramatic yellow flowers. Synthesis gas (syngas), a gas mixture that contains varying amounts of carbon monoxide (CO) and hydrogen (H_2) generated by the gasification of a carbon containing material to a gaseous product with a heating value (but less than half the energy density of natural gas). When used as a fuel, it is formed by gasification of coal or municipal waste by the subsequent reactions: $C + O_2 \rightarrow CO_2$; $CO_2 + C \rightarrow 2CO$; $C + H_2O \rightarrow CO + H_2$. The name comes from the gas's use as an intermediate in generating synthetic natural gas (SNG) and in generating ammonia or methanol. Biomass updraft gasifier, coupled with reciprocating internal combustion engines is an applied technology for minor scale power generations.

2 Updraft gasification systems

The updraft gasifier is the simplest and oldest form of the fixed bed gasifier. Syngas flow countercurrent to each other. The responsive agent is injected at the bottom of the reactor and ascends to the upper while the fuel is introduced at the top and descends to the bottom through zones of gradually increasing temperatures (drying, pyrolysis, gasification and oxidation). Heat from the gasification and oxidation zones rises upward to deliver energy for the pyrolysis and drying zones. Gases, tar and other volatile compounds are dispersed at the top of the reactor while ash is removed at the bottom. The syngas normally contains high levels of tar, which must be removed or further converted to syngas for use in applications other than direct heating. The producer gas exits at low temperature (80-300 °C) and contains an abundance of oils and tar (10-20 %) since the products of the pyrolysis and drying zone are directly drawn into it without decomposition. The dust content in the producer gas is low due to low gas velocities and filtering effect of feed in drying and pyrolysis zones [1-4]. Yang, et al.,

(2006) investigated high temperature agent gasification (HiTAG) of wood pellets in a batch type updraft fixed bed gasifier [5]. The gasifier was a cylindrical straight up reactor consisting of airstream box and gasifier bed. The feedstock bed was supported by a bed of ceramic balls placed on perforated disk within the reactor. The gasification method became more rapidly with raise in temperature of feed gas (650 to 830 ºC). The higher feed gas temperature led to increase in concentration of CO (20.1 to 26.8 %) and H_2 (6.6 to 12.7 %) in producer gas. The critical feed gas temperature (at which the yield of gaseous products was at the maximum) was between the ignition temperature of biomass and the melting temperature of the ash in the biomass. The updraft gasifier was normally used for gasification of conservative biomass fuel like wood, wood chips, bark, etc. There are many researches deal with the updraft gasifier as follows; Rao, et al., (2004) studied the gasification of unconventional fuels like low-density refuse derived fuel (RDF) pellets in an updraft gasifier with an inclined grate [6]. The grate inclination helped in crushing of large clinkers. An air partitioned in the gasifier ensured uniform distribution of air in the combustion zone. Higher heating value (HHV) and energy content of producer gas was 5.58 MJ/Nm^3 and 12.2 MJ/kg, respectively. The tar content from RDF pellets was 45 % less than that in gas generated from wood chips. The particular gasification speed of RDF pellets was within the range reported for updraft gasifiers with fixed grate i.e. SGR 100-200 kg m-2 h-1. The cold gas efficiency obtained with RDF was 73 %, which was over 8 % higher than that obtained with wood chips. Na, et al., (2003) gasified flammable dissipate pellets in an updraft fixed bed gasifier creased with alumina noncompliant using O_2 as the gasifying medium [7]. The composition of H_2 and CO in the producer gas was in the range 30-40 % and 15-30 % respectively depending on oxygen/waste ratio. At the bed height was increased the H_2 and CO content in producer gas increased whereas CO_2 content reduced. Gasification was difficult above a bed height of 700 mm because of the pressure drop of O_2. The cold gas effectiveness was approximately 61 % in the heating value of producer gas was 11.72-13.40 MJ/Nm^3, which was more than the heating value of producer gas generated by air gasification and the O_2 / waste ratio of 0.4-0.5. The gas was cleaned of the particulate matter by a cyclone. O_2 gasification was reported to generate an intermediate heating value (MHV) gas (10-18 MJ/Nm^3), which is appropriate for limited pipeline distribution and as synthesis gas for conversion to methane and methanol. The purpose of this research was to study the syngas production from Golden shower tree biomass in Klong Luang District, Pathum Thani Province for producing engine shaft power and the electricity. Charcoal wood chips converted to producer gas from the updraft gasifier stove and distributed to the gasoline engine. An experimental study was carried out on July 25-26, 2017 by using an updraft gasifier, modified 4-stroke petrol engine 13 horsepower. Finally, information on the typical performance of the gasoline engine fueled with syngas is presented.

3. Experimental setup

The experimental was step by step set up, first of all in the 13-HP gasoline engine for multipurpose used was selected to be modified for the test. The carburetor was drilled and a copper pipe gas tube for conveying producer gas to the mixing chamber was attached. The measuring equipment such as volume air flow, thermocouples, and gas flow measuring equipment were setup and recording data as shown in Figure 1-5. The power output from an engine operating on syngas was determined by the same factors as for engines operating on liquid fuels, namely: - the efficiency with which the engine changes the thermal of the flammable mixture into mechanical energy (shaft power); - the number of combustion strokes in a given time (number of revolutions per minute: rpm).

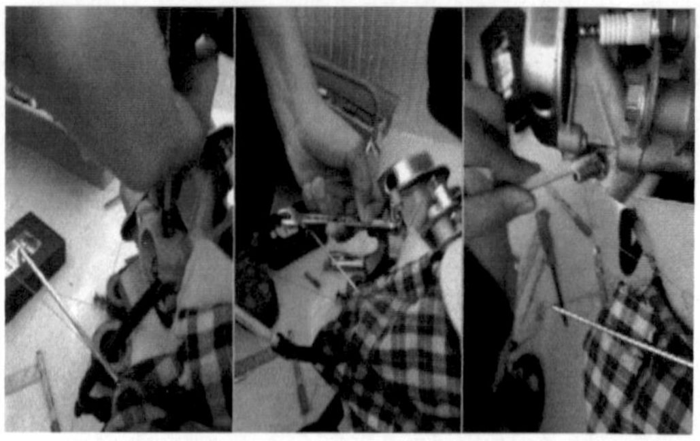

Figure 1. The carburetor of the engine was modified for the use of producer gas-air mixture.

Figure 2. Air volume, mass flow measurement of the air blower gasifier stove charger.

Figure 3. Combustion chamber of the the updraft gasifier.

https://doi.org/10.1051/matecconf/201818703002

Figure 4. The updraft gasifier experimental setup.

Figure 5. Syngas production and engine running test with the data acquisition.

The results of the engine performance test when the load on the engine requires the following calculation:
The engine shaft work is calculated from the equation.

$$P = \frac{2\pi NT}{60} \quad (1)$$

N = Engine revolution (rps)
The generator efficiency can be calculated from

$$\eta_{gen} = p_{gen} / p_e \quad (2)$$

4 Experimental results

The experimental test of the engine running used syngas from the charcoal indicated that the average engine speed was 738.33 rpm, the average power was 12.77 V, the average electric current was 6.9 A, the generated power was 88.1 W, and the engine power was 110.125 W as shown in Figure 6.

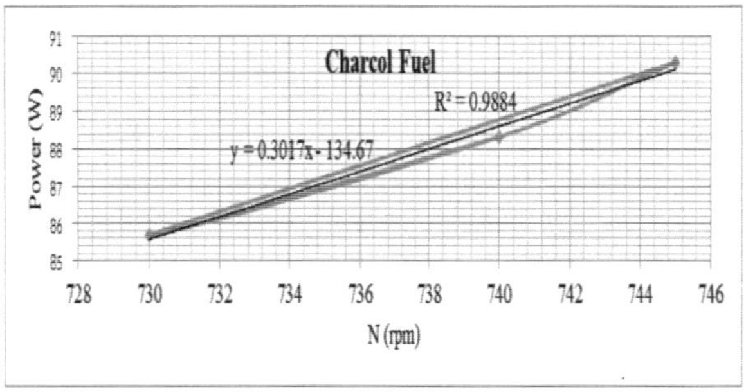

Figure 6. The relationship between the electric power and the engine speed using a charcoal fuel.

The experimental test of engine running used producer gas from the Golden Shower Tree Chips indicated that the average engine speed was 786.67 rpm, average power was 13.07 V, average power was 7.07 A, the generated power was 92.34 W, and the engine power was 115.425 W as shown in Figure 7.

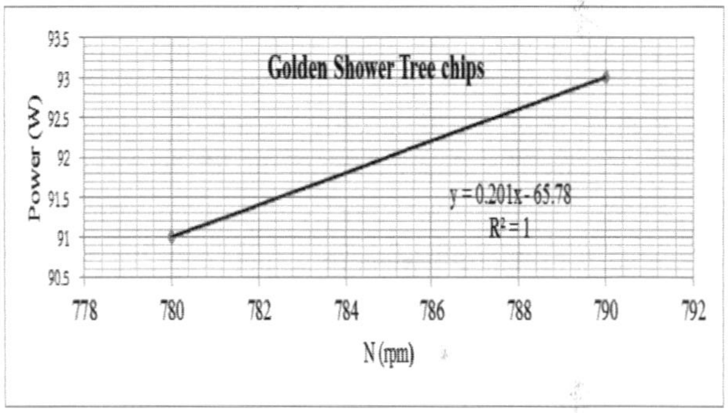

Figure 7. The relationship between electric power and the engine speed using a Golden Shower Tree chip.

The experimental test of engine running used the gasohol 91 indicated that the average engine speed was 813.33 rpm, the average power was 13.57 V, the average power was 7.55 A, the generated power was 102.43 W, and the engine power was 128.038 W as shown in Figure 8.

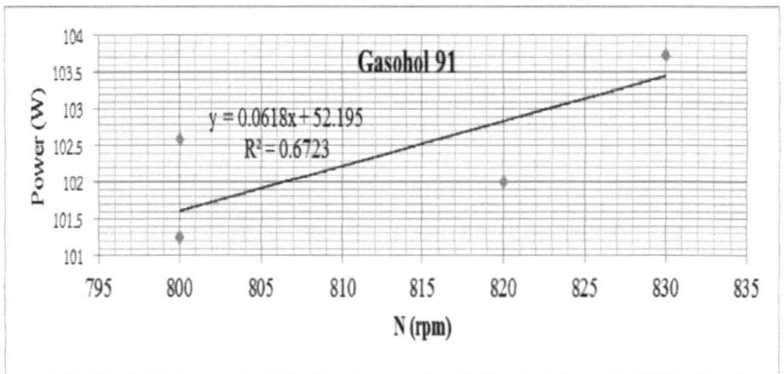

Figure 8. The relationship between the electric power and the engine speed using gasohol 91 fuel.

According to the data obtained from this study, the use of gasoline, 91 as fuel provided more engine speed. The produced electricity was higher than the use of golden shower tree wood and charcoal as fuel. Coal combustion produces less smoke than the use of golden shower tree wood. However, golden shower tree wood gives higher heat than charcoal found in higher combustion temperatures. The golden shower tree wood and charcoal can be used for small gasoline engine

5 Summary

The generated power production depends on the engine revolution. For gasoline engine the increased power is less than linear because of fluctuations in the varied efficiency factors. The extreme speed of the engines fueled by producer gas is imperfect by the combustion velocity of the combustible mixture of syngas and air. Since this speed is little as paralleled to combustible mixtures of petrol and air, the efficiency of the engine can descent affectedly if the combustion speed of the mixture and the average speed of the piston become of the same order of magnitude. In the types of engines that are presently ready-to-wear, one can expect this occurrence to occur at engine speeds of about 2500 rpm. Engines fueled by producer gas should therefore generally be operated below this speed. During the test, the heat loss was very high because there has been always a fire at the top of the fuel compartment. The use of gasohol 91 fuel the engine speed, and the produced electricity were much higher than the use of wood and charcoal as fuel. The burning of coal produces less smoke than the use of Golden shower wood chips. The Golden shower, wood chip has higher heating power than charcoal, but produces more tar. The golden shower tree wood chips can be used as fuel for small gasoline engines, but should be equipped with a gas filter to avoid engine wear.

6 References

1. Carlos, L: High temperature air/steam gasification of biomass in an updraft fixed batch type gasifier. Ph.D. Thesis. Royal Institute of Technology, *Energy Furnace and Technology*, Stockholm, Sweden. (2005)
2. Reed, T. B. and A. Das: Handbook of biomass downdraft gasifier engine systems. Colorado: *Solar Energy Research Institute*. (1988)
3. VTT: Review of Finnish biomass gasification technologies. OPET Report No. 4. *Technical research Centre of Finland*, Finland. (2002)
4. Stassen, H. E. M. and H. A. M. Knoef: Small-scale gasification systems. The Netherlands: *Biomass Technology Group*, University of Twente. (1993)
5. Yang W., A. Ponzio, C. Lucas and W. Blasiak: Performance analysis of a fixed–bed biomass gasifier using high-temperature air. *Fuel Processing Technology* **87** (2006)
6. Rao, M. S., S. P. Singha, M. S. Sodhaa, A. K. Dubey and M. Shyam: Stoichiometric, mass, energy and exergy balance analysis of countercurrent fixed-bed gasification of post-consumer residues. J. Biomass and Bioener. **27** (2004)
7. Na, J. I., S. J. Park, Y. K. Kim, J. G. Lee and J. H. Kim: Characteristics of oxygen–blown gasification for combustible waste in a fixed-bed gasifier. Applied Ener **75** (2003)

A Novel Global Warming Solution: Use of Flue Gas to Produce Urea

Zaina AlDhaheri, Saniha Aysha Ajish,Sumaita Rahman, Laila Owda, Salma AlNowahi, Omar Chaalal

College of Engineering, Chemical Engineering Department, Abu Dhabi University UAE

Abstract. Urea is a nitrogenous organic compound which is widely used as a fertilizer and in the agricultural industry. On an industrial scale, urea can be manufactured from the reaction of carbon dioxide and ammonia. The goal of this research is to design a plant that produces 46.84 tons/hr of urea from the raw materials carbon dioxide and ammonia. The quantities of carbon dioxide and ammonia consumed in the process were 37.32 tons/hr and 28. 84 tons/hr, respectively. The carbon dioxide is obtained using a sustainable approach from the waste products, the flue gas, of a nearby power plant. In the first step of the process, carbon dioxide is extracted in an absorption column that uses ammonia and stripping columns from the flue gases emitted from a power plant. This ensures that the whole production process is environmentally sustainable and contributes to the reduction of carbon dioxide that cause global warming.

1 Introduction

Carbon dioxide is the main greenhouse gas that needs to be reduced from the atmosphere. There are various technologies used to separate CO_2 from flue gas. These include chemical solvent methods, physical absorption methods, cryogenic methods, membrane systems, biological fixation, and O_2/CO_2 combustion process. Compared to the chemical plant, power plant has a large flue gas flow, different ingredients, relatively low CO_2 concentration and other characteristics. The chemical solvent techniques are generally recognized as the most effective technologies for CO_2 removal. This requires that the researchers developed a relatively low-cost, low-energy requirement CO_2 capture technology. Among the conventional CO_2 chemical removal processes, the monoethanolamine (MEA) process has been comprehensively studied and successfully used in chemical plants for CO_2 recovery in natural gas processing. Although the MEA process is a promising system for the control of CO_2

2 Urea around the globe

Urea is one of the most demanded chemical around the globe because it increases the yield of the crops which help meet the demand of adequate nourishment of the human race. Urea is most manufactured in China, where the production of urea has been over 60 million tons since 2013. India is the 2nd largest producer of urea around the globe. It manufactures about 20 million tons of urea annually from the year 2000 as of 2016, but the production of urea is expected to slightly decline by the year 2021 [5].

2.1 Urea in GCC

Middle East is the 3rd largest to manufacture urea annually [6]. It produced around 20 million tons of urea in the year 2015 and 2016 which is 100% increase from what it used to produce in the year 2000, and accounts for 39% of the world export. Due to the abundance of the raw materials in the Middle East, it takes only 100 dollars per ton to produce urea compared to United States, Russia, and China who need over 150 dollars per ton to produce urea [5, 6]. The largest producers of urea in Middle East are Qatar, Saudi Arabia, and Iran. Qatar produced over 5 million tons of urea according to a data in 2013. On the other hand, Saudi Arabia, who is also the largest urea producer in GCC, has produced around 3.5 million tons of urea in 2013 [6]. The other GCC countries that produce urea in decreasing rate are United Arab Emirates, Oman, Kuwait, and Bahrain respectively [7].

2.2 Saudi Arabia

Saudi Arabia manufactures the largest quantities of urea and fertilizers in GCC. The production capacity of this country accounts for 36% of GCCs' total fertilizer industry. It reached about 10.9 million tons in 2012. This has been majorly contributed by Ma'aden Phosphate Company's Fertilizer Complex [7] "With the launch of Safco 5 this year, notable for possessing one of the world's largest carbon capture units, and Maaden's Waad Al Shamal phosphates facility in 2016, Saudi fertilizers will continue to see much dynamism in the near future," continued Dr. Al- Sadoun [8].

2.3 United Arab Emirates

The 2nd largest producer of urea in GCC is United Arab Emirates, which produced around 1.5 million tons of urea in the year of 2013 [6]. There are a handful of companies that produce like ADFERT, Prompt Point fze, and FERTIL (largest urea producing company in UAE) [9]. FERTIL is a joint venture between ADNOC, with 66.6% of stake, and French oil and gas major Total, with the remaining 33.4% of the stake [10]. FERTIL produces 450,000 tons per annum ammonia of which 350,000 tons per annum is utilized for the production of urea. "Fertil's yearly exports currently stand at 100,000 tpa of liquid ammonia and more than 625,000 tpa of urea (bulk and bagged)," said Yousef Omeir Bin Yousef, CEO of Adnoc [11]. This project has commissioned to expand their project to FERTIL-2, which was built at the cost of 1.2 billion dollars. FERTL-2 can produce 3,500 tons/day of urea at fullest capacity, which has greatly boosted the production of urea in UAE [10]. FERTIL exports majority of their manufacture to foreign countries like Brazil, Pakistan, Australia, and the United States [10, 12].

3 Carbon dioxide affects global warming

Since 1988, the world's environmental scientists on the Intergovernmental Panel on Climate Change (IPCC, 2014) have provided multiple and clear scientific reports on the state of knowledge on climate change. It is an undeniable fact that our planet is warming and the climate is changing. Because of the impact of global warming on our future, it is the most important environmental issue and a subject of actuality in present. For instance, extreme weather disasters have partially or completely damaged crop production in many parts of the world [13-17].

Global warming is the gradual noticeable increase in the earth's average overall temperature due to the earth's heat being trapped in the atmosphere. It is mainly caused by the greenhouse effect. One of the main gases that cause global warming is carbon dioxide. Carbon dioxide is either released into the atmosphere by natural processes like volcanic eruptions and naturally accruing fires or by human processes like the burning of fossil fuels leading to very drastic effects on the environment [18]. Power plants, oil refineries, and other factories produce CO_2 in large quantities which is beyond the earth's capacity to absorb and neutralize. Mainly in GCC countries, where the demand for desalinated water is high, the emissions of carbon dioxide are increasing rapidly. In order to decrease those environmental effects, the plant presented in this study will reuse the carbon dioxide from power plants in desalination plants in production of urea. This will reduce the overall amount of carbon dioxide emitted into the atmosphere and the plant will save up costs of producing or extracting carbon dioxide in order to be used. Recent project was studied to reduce CO_2 from either natural gas or exhaust gases where the experimental results showed excellent removal of CO_2 [19-21]. There are several ways to reduce carbon dioxide while making good use of it when dealing with reactions such as the production of urea. Moreover, putting the environmental effects under consideration while making a plant will help to reduce the effects of global warming which vary from increase of global temperature and sea levels to rough heat waves and intense hurricanes [22, 23].

4 Theory of the process: Ammonia and CO_2 Reactions

The commercial synthesis of urea involves the combination of ammonia and carbon dioxide at high pressure to form ammonium carbamate which is subsequently dehydrated by the application of heat to form urea and water. NH3 and CO2 are converted to urea via ammonium carbamate at a pressure of approximately 140 bar and a temperature of 180-185°C.

Ammonium carbamate

$$2NH_3\,(g) + CO_2\,(g) \rightleftarrows NH_2COONH_4\,(s) \qquad (1)$$

Urea

$$NH_2COONH_4 \rightleftarrows NH_2CONH_2\,(l) + H_2O\,(g) \quad (2)$$

Equation 1 is fast and exothermic and essentially goes to completion under the reaction conditions used industrially. Equation 2 is slow, endothermic, and does not achieve full conversion (about 50-80% on CO_2 basis). The conversion can be increased by increasing the temperature and the NH_3/CO_2 ratio, however, it decreases with increasing H_2O/CO_2 ratio.

The total reaction of aqueous ammonia absorb carbon dioxide can be described as in Equation 1:

$$CO_2\,(g) + NH_3\,(aq) + H_2O\,(l) \rightleftarrows NH_4HCO_3\,(aq) \quad (3)$$

The actual process of the reaction is more omplicated, which can be described as step-by-step reactions.

First of all, in Equation 2, CO_2 and NH_3 react to generate NH_2COONH_4, then NH_2COONH_4 hydrolyzes in solution instantaneously.

$$CO_2\,(g)+2NH_3(aq) \rightarrow NH_2COO\text{-}(aq)+NH_4\text{+}(aq) \quad (4)$$

Then, NH_4+ and NH_2COO- have an irreversible reaction (Equation 3) in solution:

$$NH_2COO\text{-}(aq) + NH_4\text{+}\,(aq) + 2H2O \rightarrow NH_4HCO_3(aq) + NH_3(g) + H_2O(aq) \qquad (5)$$

At the same time, the complex balances of solute ionizing and ion reactions are happening in the solution shown below:

$$NH_3(aq) + H_2O(l) \rightleftarrows NH_4\text{+}(aq)+ OH\text{-}\,(aq) \quad (6)$$

$$NH_4HCO_3(aq) \rightleftarrows NH_4\text{+}(aq) + HCO_3\text{-}(aq)$$

[Equation 7]

$$(NH4)_2\,CO_3(aq) \rightleftarrows 2NH_4\text{+}\,(aq)\text{+}CO32\text{+}(aq)$$

[Equation 8]

$$OH\text{-}(aq) + HCO_3\text{-}(aq) \rightleftarrows CO32\text{-}\,(aq) + H_2O \;(9)$$

$$CO32\text{-}(aq) + CO_2\,(g) + H_2O(l) \rightleftarrows 2HCO \quad (10)$$

Presentation of the process:

The proposed process is done in two steps and the whole process is presented in Figure 1.

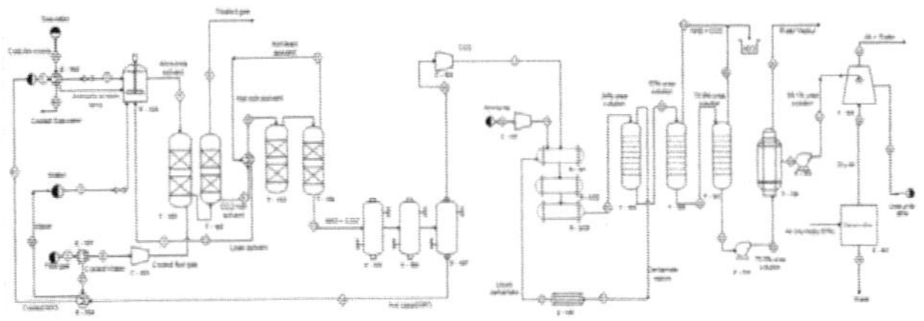

Figure 1. Overall process flow diagram Step 1.

Step 1: Carbon dioxide capture and separation

The first step of the process, which is the sustainable approach, in this project is to isolate carbon dioxide gas from the flue gases emitted from a nearby power plant, which is done by absorption using ammonia solvent. In order to prepare the ammonia solvent, liquid ammonia is purchased from a supplier and stored in the vicinity of the plant. A valve controls the flow of this liquid ammonia to a mixer where water is also added to make a 30% ammonia solution to be used as a solvent. Since the temperature of the stored ammonia is extremely low, it is made to pass through a heat exchanger initially where the temperature is raised to room temperature.

The hot flue gas obtained from the nearby power plant cannot be used at high temperatures. In order to bring down the temperature, it is passed through a cooler where the water vapor gets condensed and is recycled back to the water source to prepare the ammonia solution. Once the gasses are cooled, a blower push them to an absorption column that operates at atmospheric pressure and normal temperature.

In the absorption column, the flue gas enters from the bottom and the ammonia solvent enters from the top. The flow is countercurrent where the treated gas, mainly nitrogen, is left to the atmosphere while the carbon dioxide rich solvent is made to enter a heat exchanger. Once the carbon dioxide rich solvent gets heated, it is made to enter a regenerator where two outlet streams are obtained: hot lean solvent and another containing ammonia and carbon dioxide.

In order to obtain the carbon dioxide alone, the former outlet stream from the regenerator is sent to a liquid separator where the carbon dioxide rises to the top of the column and exits out. It is then directed to the second step of the process to be used for urea manufacture. The ammonia liquid which exits from the bottom of the separator gets recycled back to the mixer to make more of the ammonia solvent to be used for absorption. It is important be note that once the process reached steady state, the valve that controls flow of ammonia from the storage tank is closed so that the ammonia solvent keeps getting recycled in the closed loop. This ensures that the process is more profitable and sustainable.

Step 2: Production of urea

The second part of the process is to actually manufacture the urea. This occurs in a reactor which is at high temperature and high pressure ensuring maximum conversion of the reactants. The reactants used will be the carbon dioxide obtained from the previous process as well ammonia gas which is pumped into the reactor. The end of the reaction shows that 34% urea solution is produced along with other undesired products such as water, ammonia and carbamate. In order to purify the urea more, it sent to a stripping column where the carbamate separates out from overhead, The carbamate is then sent to a carbamate condenser where it is condensed back to liquid from and then recycled back to the reactor.

The rest of the 67% urea solution is sent to a decomposer at medium pressure. Here, 50% of the undesired ammonia and carbon dioxide gases are removed from the top while giving 70.9% pure urea solution. This is then sent to another decomposer at low pressure where the rest of the ammonia and carbon dioxide are removed and the only remaining solution 75.3% urea.

In order to remove the water from this solution, it is sent to an evaporator at vacuum conditions and low pressure, where water gets evaporated and is let into the atmosphere. The remaining 95.1% urea solution is sent to a prilling tower to further remove moisture content.

In the prilling tower, dry air is required to carry out the process. This dry air is obtained after passing humid air to a dehumidification unit. The urea solution is sprayed in the prilling tower where it counteracts with the dry air to form urea prills of 96% purity. This is the end product of the urea manufacture which can then later be stored and sold.

5 Conclusion

The research in this paper has been developed using the most practical but cost efficient and sustainable measures to produce urea. One of the main reactants needed for this production, carbon dioxide is obtained through an environment friendly approach. Being located next to a power plant, waste flue gases rich in carbon dioxide can easily be collected and used to produce urea. Using this innovative technique, carbon dioxide gas is obtained from the waste product which would have otherwise been emitted into the atmosphere to cause global warming. The flue gases undergo a series of absorption and regeneration processes with ammonia solvent to get the pure carbon dioxide. The urea is manufactured using this innovative process to get the final commercial product – solid urea prills. The whole process has been developed using the most efficient sustainable approach.

References

1. Chakma, A. "Separation of CO_2 and SO_2 from Flue Gas Streams by Liquid Membranes". Energy Convers. Manage. 1995, 36, 405.
2. Molburg, J. C.; Thimmapuram, P.; Doctor, R. D. Comparison of Salient Features of Alternatives CO_2 Recovery Processes. A&WMA Annual Meeting, Cincinnati, June 1994; Paper 94-RA113.02.
3. Zhiming Zhang et al., "Research on Physical Chemistry Characters and Production Increasing

Mechanism of Long Effective Ammonium Bicarbonate" China Science (B), 26, 453 (1996)
4. Bai HL et al., Removal of CO_2 Greenhouse Gas by Ammonia Scrubbing, Ind Eng Chem Res, 36, 2490 (1997)
5. Y. Yamaguchi, "A world awash in urea", in 32nd Annual World Petrochemical Conference, 2017.
6. "Home - Industry Overview - PotashCorp", Potashcorp.com, 2017. [Online]. Available: http://www.potashcorp.com/overview/nutrients/nitrogen/overview/global-urea-profile. [Accessed: 23-Oct-2017].
7. 2017. [Online]. Available: http://www.gpca.org.ae/ff2012/53.pdf. [Accessed: 23- Oct- 2017].
8. "GCC fertilizer industry more than doubles in the last decade, says GPCA", Gulf Petrochemicals & Chemicals Association, 2017. [Online]. Available: http://www.gpca.org.ae/news/gcc-fertilizer-industry-more-than-doubles-in-the-last-decade-says-gpca/. [Accessed: 23- Oct- 2017].
9. "List of Urea Fertilizer companies in United Arab Emirates", Listofcompaniesin.com, 2017. [Online]. Available: http://www.listofcompaniesin.com/united-arab-emirates/urea-fertilizer/. [Accessed: 24- Oct-2017].
10. D. Thapliyal, "UAE's FERTIL to start urea, ammonia production at new plant in June", Icis.com, 2013. [Online]. Available: https://www.icis.com/resources/news/2013/05/13/9667976/uae-s-fertil-to-start-urea-ammonia-production-at-new-plant-in-june/. [Accessed: 24- Oct- 2017].
11. S. Carvalho, "Fertil joins hands with France's Total to enhance urea production capacity", Gulf News, 2006.
12. "2013 Minerals Yearbook", 2015. [Online]. Available: https://minerals.usgs.gov/minerals/pubs/country/2013/myb3-2013-ae.pdf. [Accessed: 24- Oct- 2017].
13. Battisti, D. S. & Naylor, R. L. Historical warnings of future food insecurity with unprecedented seasonal heat. Science 323, 240–244 (2009)
14. World Food Program . Pakistan – Flood Impact Assessment http://documents.wfp.org/stellent/groups/public/documents/ena/wfp225987.pdf (2010)
15. Gu, L. et al. The 2007 eastern US spring freeze: increased cold damage in a warming world. Bioscience 58, 253 (2008)
16. Barriopedro, D., Fischer, E. M., Luterbacher, J., Trigo, R. M. & García-Herrera, R. The hot summer of 2010: redrawing the temperature record map of Europe. Science 332, 220–224 (2011)
17. Coumou, D. & Rahmstorf, S. A decade of weather extremes. Nature Clim. Change 2, 491–496 (2012).
18. "Cite a Website - Cite This For Me", Climate.nasa.gov, 2017. [Online]. Available: https://climate.nasa.gov/effects/. [Accessed: 22- Oct-2017].
19. O. Chaalal and M. Hossain, "A two-step process for removal of carbon dioxide and production of desalinated water using ammoniated saline water," Environmental Progress & Sustainable Energy Volume 35, Issue 1, Version of Record online: 3 JUL 2015.
20. Hameed Muhamad and Omar Chaalal "Optimization Study for Carbon Dioxide in flue gas using liquid Membrane Permeator". International Conference: The 9th Dynamical Systems & Applications July 20-23, 2016 Antalya, Turkey
21. Omar Chaalal, Hameed Muhamad, Md Monwar Hossain and Dalila Hank "A numerical study in a carbon dioxide membrane permeator " Far East Journal of Mathematical Sciences (FJMS) Volume 99, Issue 12, Pages 1833 - 1847 (June 2016)
22. "Global Warming Effects", Nationalgeographic.com, 2017. [Online]. Available: https://www.nationalgeographic.com/environment/global-warming/global-warming-effects/. [Accessed: 22- Oct- 2017].
23. R. Jackson, "Global Climate Change: Effects", Climate Change: Vital Signs of the Planet, 2017. [Online]. Available: https://climate.nasa.gov/effects/. [Accessed: 22- Oct- 2017].

Linear Programming Model for the Measurement of Environmental Performance in Wine Industry—Concepts and Empirical Results

Shiang Chen and Jung-Hua Wu

Department of Resources Engineering, National Cheng Kung University, 70145 Tainan City, Taiwan

Abstract. In Taiwan we are currently facing serious air pollution problems, PM2.5 airborne particles not only endanger the human body, leading to coughing, asthma, chronic bronchial and other respiratory diseases, but it is also carcinogenic, our basic rights for survival has been seriously threatened. According to the latest study by the Environmental Protection Agency, the industrial sector accounts for about 25% of the sources of air pollution in Taiwan. The winemaking industry, with raw materials has several stages of production; it must use electricity, water, oil, diesel, gasoline, and steam as energy. The final product is not only wine, but also the discharge wastewater, lees, CO_2, waste gas, etc. Different process methods will produce different energy consumption results, thus providing the possibility of CO_2 reduction. This study takes the winemaking industry as a case study, trying to find out how to achieve a concept of "win-win situation "for both sides of carbon reduction and economy development. First, aiming at the policy of energy saving and carbon-reduction inventory of winery, trying to find out the source of pollution and the change of pollution source during 2008-2016, and then according to the data, the empirical results are analysed. Finally, a linear programming model proposed for the production planning of winemaking processes to achieve maximum operational profit while reducing CO_2 emissions. The options considered in this research are levy the carbon tax in different prices to decrease CO_2 emissions; installation of new equipment to enhance capacity; switching production schedule that emits less CO_2. The objective of the linear programming model is to determine suitable CO_2 mitigation options for a given reduction target while meeting the demand of each final product, quality specifications, and simultaneously maximizing profit. In this research, How to achieve the optimal production goals: minimizing the cost & energy consumption, and the maximum production value become the optimal production objection.

1 Introduction

In Taiwan, trading market of CO_2 emission is small, the effect of carbon tax maybe better than emissions trading. Therefore, we use linear programming, carbon tax, and inventory of energy saving to explore optimal production in the winemaking industry. This paper presents linear programming for the optimal production, it based on the input–output activity to assess the potential impacts of energy consumption, CO_2 emission and seek a production schedule that will increase the profit of winemaking industry [1]. There are seven kinds of product for research object; red label rice cooking wine (code a1); red label pure rice cooking wine (code a2) ; super red label rice wine (code a3); super red label pure rice wine (code a4); super red label pure rice wine (code a5) ; red label rice wine water (code a6); wine gift box (code a7). Data are analysed by MATLAB simulation software, and use CVX module to solve problems. This optimization model used to estimate the potential CO_2 emissions mitigation strategies by minimizing the CO_2 emissions under the different prices of carbon tax. Several mitigation strategies have applied to analyse the effect of CO_2 emissions reduction. In this paper, first, we describe the data source and methodology, then the linear programming model of the winemaking industry established, aiming at the goal of minimizing costs and maximizing profits, we discuss the influence of changes to energy consumption and the decrease of CO_2 due to the levy of carbon tax [2]. Finally, a case study of greenhouse gas inventory for energy saving and carbon reduction in winery are analysed. We analyse and compare the sources of greenhouse gas emissions in the winery during 2008-2016.

2 Data source and methodology

Carbon reduction projects in Taiwan has been in operation for many years, according to the figure 1, the latest carbon emissions results released in 2017, the overall carbon emission is the highest in nearly a decade. Our greenhouse gas reduction targets are to achieve greenhouse gas emissions reductions from 2.93million /ton reduce to 2.82 million/ton in 2020. This paper attempts to option the effective policy that can assist Taiwan to reduce the CO_2 emissions. All the data are provided by winery, it's an actual operating condition in 2017, and an optimization model" linear programming model" is to estimate the potential CO_2 emissions mitigation strategies for winemaking industry by minimizing the CO_2 emissions under the constraint of different prices of carbon tax [3].

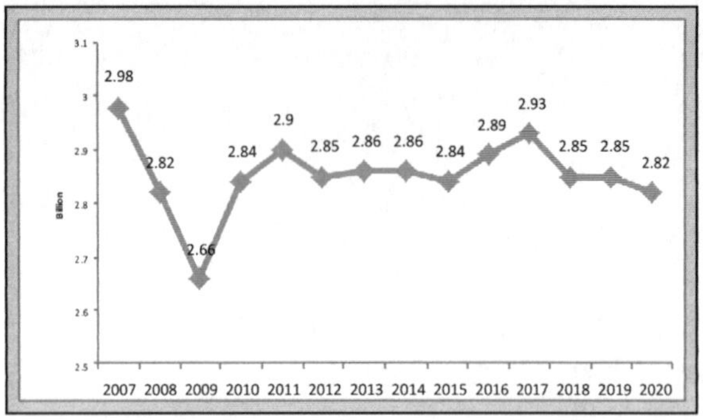

Figure 1. CO_2 Emission Trends Overall Taiwan.
Source: http://teds.epa.gov.tw

2.1 Definition of variables

2.1.1 Formatting the title

a_{ij} :The input of i process to the j item process, and the decision variable for the research model

C_{jp}: Cost refers to the wine system on the j process

p_j: Price factor of j product

$\overline{CO_2}$: Upper constrains of the total CO_2 emission.

e_{qj}: The j production process, the need for q-type energy unit input

Pc: It indicates the CO_2 tax rate

u_q: CO_2 emissions caused by the use of the q energy

$$\underset{a_i \ldots a_M}{maximize} \; \sum_{i=1}^{M} P_i a_i - \sum_{i=1}^{M} C_i\, a_i - \sum_{i=1}^{M} E_i\, a_i - P_c \sum_{i=1}^{M} Q_i a_i$$

$$\text{subject to } a_i >= 0, \; i = 1, \ldots, M$$

$$\sum_{i=1}^{M} a_i \le CAP$$

$$a_i \le \bar{a}_i \quad i = 1, \ldots, M.$$

$$\sum_{i=1}^{M} Q_i a_i \le CO_2 \qquad (1)$$

Where M is 7, i.e, 7 products

a_i Means the quantity of product produced in one month.

P_i Means the price of product

C_i Means the cost of product

E_i Means the energy consumption (water, fuel, and electricity) of product

P_c Means the carbon tax

Q_i Means the CO_2 emission of product

CAP Means total quantity limitation

$\bar{a}_i$ Means the limitation quantity of product

$CO2$ Means limitation of CO_2 emission

eq(1) means each quantity of product should be positive.

eq(2) means total quantity should less than total quantity limitation.

eq(3) means the quantity of product should not larger than the limitation of it.

eq(4) means the emission should not larger than limitation of CO_2 emission.

3 Case study of winery

In Taiwan, our greenhouse gas reduction targets are to achieve greenhouse gas emissions reductions for 2025, and to encourage energy-intensive industries to use the best energy efficient technology to effectively manage energy and improve efficiency, especially in industries that must seek the most efficient usage of energy, that is, the minimum CO_2 emissions operation, to pursuit profit maximization. According to Figure 2, the energy consumption in the industrial sector is the largest, in 2016, for example, the energy consumption of the industrial sector accounts for about one-third (37.2%) of the total consumption of energy. Overall, the industrial sector is the most important energy consumer, and the main source of CO_2 emission [4].

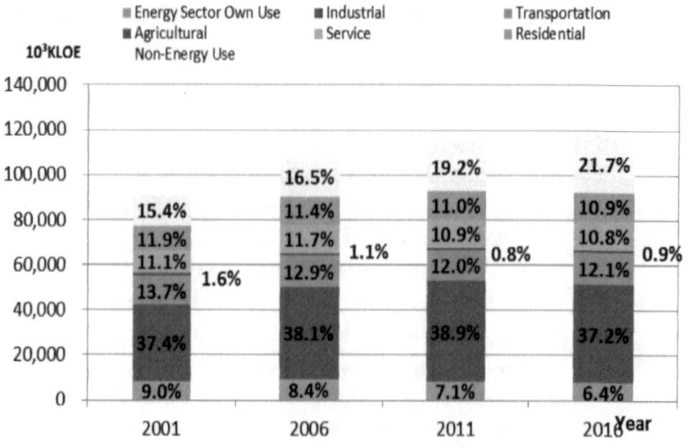

Figure 2. Taiwan's Energy Supply Structure (Sector).

Bureau of Energy (2016)

Refer to the Figure 3, the winemaking industry, with rice, millet, barley, or sorghum as raw materials, has several stages of production process, such as soaking, cooking, cooling, fermentation, distillation, storage, blending, filtration, bottling, packaging, and transportation. During the process distilleries must use electricity, water, oil, diesel, gasoline, and steam as energy. The final product is not only wine, but also the discharge wastewater, lees, CO_2, waste gas, etc. Different process methods will produce different energy consumption results, thus providing the possibility of CO_2 reduction. As the winemaking industry maximizes the pursuit of profits, so in this study, first we must establish the linear programming model of the winemaking industry, then under the goal of profit maximization, we explore the carbon tax and the interrelationship caused by changes in energy consumption and the impact of CO_2 reduction, finally we analyze the CO_2 inventory policy of the winery during a period from 2008 to 2016.

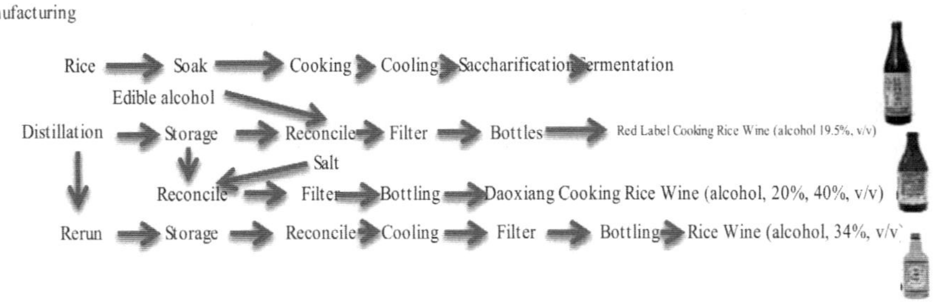

Figure 3. Process of wine making.
source: designed by the author

3.1. The policy of greenhouse gas inventory in winery

The organization of greenhouse gas emissions inventory established on January 6, 2005 and divided into many departments. The classification and definition of emission sources according to ISO/CNS 14064-1 and GHG Protocol are as follows:

Category 1: Direct greenhouse gas emissions mainly caused by routine operating activities.

Category2: Indirect greenhouse gas emissions (electricity).

Category 3: The other indirect greenhouse gas emissions.

With the policy of energy saving and carbon reduction measures, winery actively committed to the greenhouse gas inventory from 2010, they held regular meetings, scheduled time and audit date since 2012. Analysis of the types of greenhouse gas emissions from 2008 to 2016, as shown in Figure 4, the source of fixed fuel oil as the main source, followed by wastewater, process emissions, electricity and outsourcing. Figure 4 shows that fuel oil, electricity, outsourcing and fixed diesel CO_2 Emissions reduced significantly, especially in the other escape emission and acetylene reduction, but the wastewater, refrigerant escape and septic tank CO_2 emissions increased a bit.

Unit: metric ton co_2/year

3.2 Greenhouse gas emission sources in winery 2008-2016

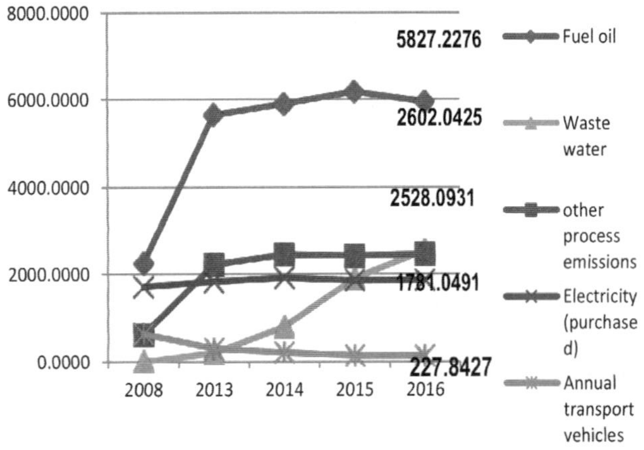

Figure 4. Greenhouse Gas Emission Sources 2008- 2016

3.3 Comparison of emissions source during the period 2015-2016

From Figure 5 comparisons of emissions 2015-2016. Fuel oil decrease from 49% to 45%, the process emissions decrease from 20% to 19%, electricity maintains 14%, wastewater increase from 14% to 20%. Wastewater discharge is still a problem, due to mainly increase in production; winery must to help the distiller's grains sales and removal to solve the problem.

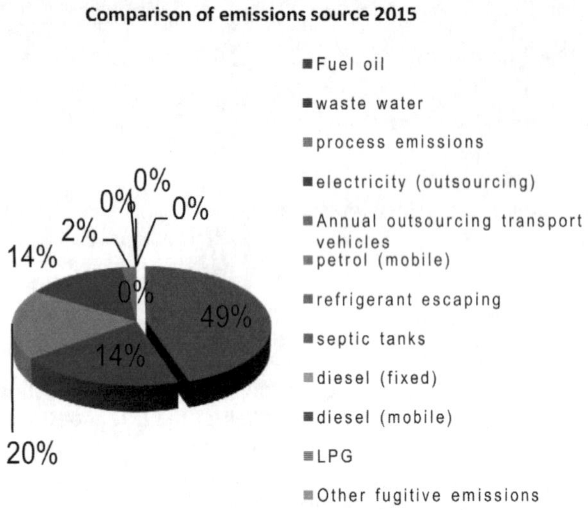

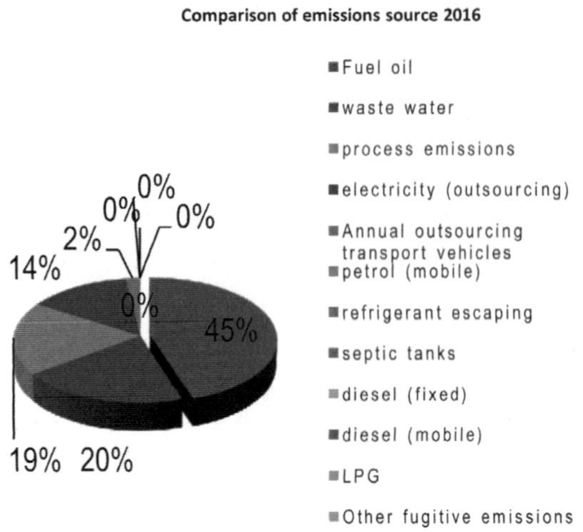

Figure 5. Emissions Source of 2015-2016.
Source: Author draws on data

3.4 Comparison of energy intensity.2005-2016

Energy intensity is an international indicator of the overall energy efficiency of a country. It represents the production process, the output of each unit of domestic gross domestic product (GDP) need to consume energy, calculated by the total energy consumption, it divided by gross domestic product.
Energy Intensity = Energy Consumption / GDP

If a country has a high energy intensity, it means that the cost of energy conversion to GDP is high; if the energy intensity decline will show that the creation of a unit of gross domestic production, the amount of resources usage is declining, the efficiency energy usage has improved , the situations of environment damage is also reduced. From Figure 6 Intensity Ratio, It shows that product intensity, service intensity and turnover intensity in the period 2005-2016 are trend of decreasing and flat for a long-term.

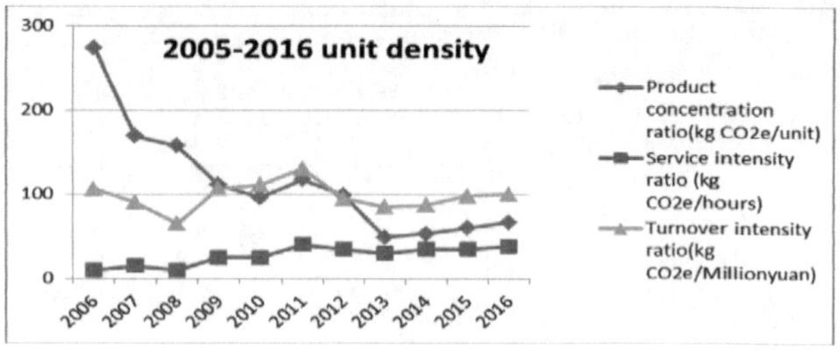

Figure 6. Comparisons of Energy Intensity 2005-2016.
Source: Author draws based on data.

4 Linear programming model for the measurement of environmental performance

In this research, first we establish linear programming model then set carbon tax prices, by CO_2 tax mechanism to analyse changes in the production scheduling process and energy usage, finally under the goal of maximizing profit, to explore the interaction between changes in production scheduling and CO_2 emissions reductions.

4.1 The goal of the linear programming model

The goal of the linear planning model of wine industry is the combination of output interest, input cost and carbon cost, expression in monetary terms; the energy goal is to minimize energy input, in terms of quantity or currency; the environmental goal is to minimize CO_2 emissions by quantity. According to the Figure 7, there are three targets of the linear planning model. They are economy, energy and environmental protection [5]. Since the purpose is to produce in the production process of obtaining the maximum profit, however, profit is the value of the product portfolio produced in the production process and the cost invested in the manufacturing process, therefore in this study; we must consider minimizing CO_2 emissions the amount of environmental objectives. The following chart shows the concept of wine industry; its targets are as follows [6]:

1. Economic target:

Optimize the allocation of resources, the pursuit of the maximization of production value. The mathematical formula be measured in monetary units

$$\text{Max}\sum_{j=m+1}^{n}\sum_{i=1}^{m} p_{ja_{ij}} - \sum_{j=1}^{m}\sum_{i=1}^{m} c_{jpaij} \quad (2)$$

2. Energy Target:

Fixed capital investment, the pursuit of the minimum energy supply

$$\text{Energy goal: } \sum_{j=1}^{m}\sum_{i=1}^{m} p_{cQ_j}\, a_{ij} \quad (3)$$

3. Environmental target:

The pursuit of optimal profit conditions, CO_2 emissions are minimized

$$\text{Environmental goal: Min}\sum_{j=1}^{m}\ \sum_{i=1}^{m} Qjaij \quad (4)$$

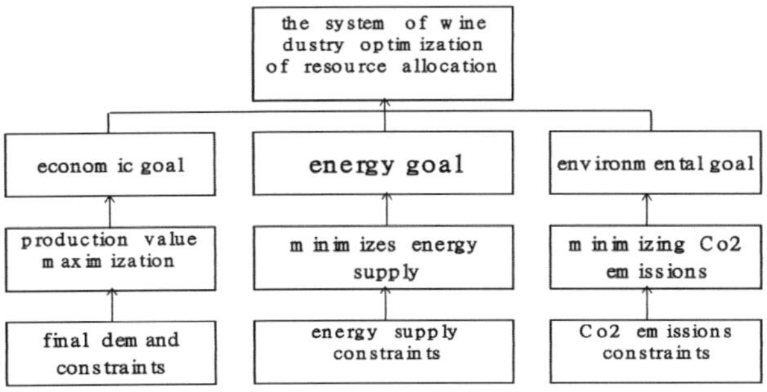

Figure 7. The Goals of Linear Programming.
Source: Author draws based on data

4.2 Application of linear programming model

The purpose of this study is to explore the impact of the levy of carbon taxes on CO_2, emissions and to pursue environmental goals with minimal CO_2, emissions. First, we give a tax level to explore the impact of the levy of carbon tax, then multiply the emissions by the tax rate and convert it into currency units. The environmental goal is mainly that when the carbon tax occurs CO_2 emissions can achieve a minimum goal [7] so this study sets the minimum environmental cost in the target equation. It is as follows:

$$\text{Min} \sum_{j=1}^{m} \quad \sum_{i=1}^{m} Qjaij \tag{5}$$

At different levels of carbon tax, the changes in production schedules within each process of the winemaking framework, the changes in energy consumption and the impact on CO_2 reductions both discussed. To maximize the value of production under the goal of minimizing the use of energy demand and minimizing CO_2 emissions.

4.3 Empirical simulation results

The application of linear programming model, first we embody the model, and then use Mat lab software to simulate the changes in the production schedule, which based on the level of carbon tax and under the basic situation of minimizing energy usage. Refer to the Table 1, the results show that the production benefit is (6731809) U.S. dollars per month, the total energy consumption is (12342.45) Kcal/month, and the total CO_2 emission is (1743.714) metric ton/month. At this time, CO_2 has not set price, it regard as the total emission limitation, so in this study we use this total CO_2 emission as the CO_2 emission constrain.

Table 1. Empirical simulation results under different carbon taxes.

CO2 price	0	90	120	150	180	210	230
CO2	**mt**	**/month**					
Total emission	52311.42	26824.93	26824.91	23845.76	23845.76	18082.02	17714.83
fuel	12342.45	12342.48	12342.48	12331.86	12331.86	2331.86	12331.86
electricity	83.87539	83.86877	83.86877	83.82213	83.82213	83.80743	83.81134
steam	12248.64	12247.71	12247.71	12241.07	12241.07	12239.59	12240.17
cost	**US$**	**/month**					
Total	3516391	3150450	3686948	4302405	5017778	4797977	5288715
Fix cost	1982900	1854641	1854641	1765104	1765104	1569835	1573156
Variable cost	1533491	1027559	1027559	1106556	1106556	1058300	1058335
CO_2 cost	0	268249.3	804747.4	1430746	2146119	2169842	2657225
value	6731809	5850126	5850125	5696954	5696954	4898471	4857764

According to the Figure 8, different carbon tax prices have different ways of distributing production capacity: When carbon taxes levels are low, wine producers will continue to produce high-energy but with a high profits product.

According to figure 9 while the carbon tax is higher, the industry will not produce products with high carbon emissions; instead, they will choose products with lower net profit, but lower energy consumption.

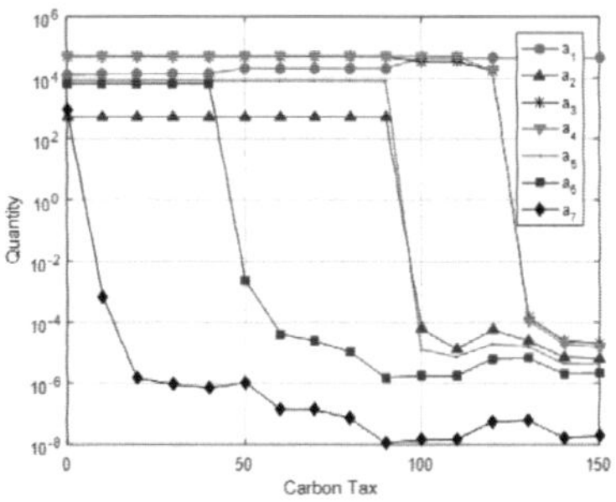

Unit: us$/Mt

Figure 8. The Interaction between Carbon Tax and the Quantity.

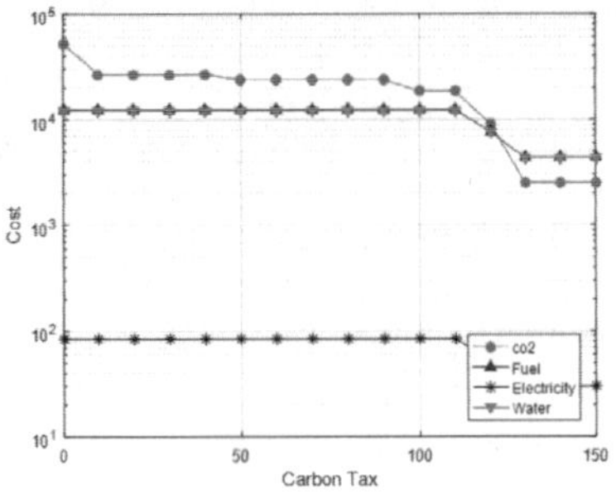

Unit: us$/Mt

Figure 9. The Interaction between Carbon Tax and the Cost of Energy.

4.3.1 Cost changes at different carbon tax levels

After the levy of the carbon tax, the winemaking producer will reduce the fixed cost and operating cost through reducing the use of fuel and adjusting the operating cost to reduce the cost of CO_2 emissions. Refer to the Table 2 owing to the tax costs of CO_2 emissions, the total cost will increase. However, as the carbon tax level increases, the ratio of fixed costs and operating costs to total costs will decrease slightly.

Table 2. Cost changes at different carbon tax levels.

CO_2	0	90	120	150	180	210	230
Total cost	3516391	3150450	3686948	4302405	5017778	4797977	5288715
Fixed and operating costs	3516391	2882201	2882200	2871660	2871660	2628135	2631490
%	100	91.48537	78.17307	66.74544	57.2297	54.7759	49.75671
Tax costs	0	268249.3	804747.4	1430746	2146119	2169842	2657225
%	0	8.514633	21.82693	33.25456	42.7703	45.2241	50.24329

4.3.2 Tax effect analysis

In terms of the concept of cost-effectiveness, the best carbon tax effect can obtain when the cost of CO_2 emission reduction is equal to the cost of carbon tax. At this point, the cost of investing in CO_2 to reduce emissions will be equal to the taxable cost of the CO_2 being collected [8]. In other word, the relative CO_2 emission at this time is the optimal production level. According to Table 3 when the cost of reducing CO_2 emissions is lower than the cost of tax, the winemaking producer will try its best to reduce CO_2 emissions. However, when the cost of reducing CO_2 emissions is higher than the cost of tax, winemaking producer will adopt a strategy of preferring to pay carbon tax to save costs. That is to say, after the CO_2 tax, whether the producer will substantially reduce its carbon emissions will determine whether the taxable cost of CO_2 has already exceeded the CO_2 emission reduction costs. When CO_2's taxable costs are lower, it is less likely to commit to carbon reduction because carbon tax is beneficial to the industry. In other words, after the CO_2 tax, whether the wine industry will actually reduce CO_2 emissions decide whether the tax cost of CO_2 has already exceeded the CO_2 emission reduction cost. When the CO_2 tax cost is low, because the payment of carbon tax is beneficial to the industry, it is less likely to commit itself to commit carbon reduction [9].

Table 3. Tax effects under different carbon taxes.

CO2 price US/Mt	90	120	150	180	210	230
CO2 reduction						
Mt /month	25486.49	25486.5	28465.65	28465.65	34229.4	4596.59
%	48.90861	48.90862	55.98457	55.98457	72.29255	72.99447
Cost of Profit and loss point US/Mt	39251.27	39251.26	36261.44	36261.44	30497.68	30130.5
CO_2 Tax effect						
US/Mt	26824.93	26824.91	23845.76	23845.76	18082.02	17714.83
The proportion of carbon tax%	68.34154	68.34153	65.76066	65.76066	59.2898	58.79368

5. Conclusion

In this research, we make use of linear programing model to explore complicated winemaking process architecture and compare changes in CO_2 emissions before and after the carbon tax levied. The process of simulation is first to simulate the situation of carbon tax which has not been levied as a basic situation, and to calculate the total emission of CO_2 under the basic situation simulation as the discharge constrain, then to observe the interaction between the carbon tax level and production capacity. Due to our goals are to minimize the demand for energy usage; to minimize CO_2 emission, and pursuit the maximizing production value.

According to the tax effect analysis, the result shows that:

1. Different carbon tax level has different ways of distributing production capacity.
2. When carbon tax levels are lower, wine producers will continue to produce high energy but with a high profits product.
3. While the carbon tax is higher, the industry will not produce products with high carbon emissions. Instead, they will choose products with lower net profit, but lower energy consumption, in other words the wine industry will undergo to change production schedules that will change the usage of energy and thus reduce the total CO_2 emissions. Especially the huge enterprises only high level of carbon tax can force them to reduce the CO_2 emission.
4. After carbon tax levy, wine industry is under pressure to reduce their tax costs and thus change their production schedules, and by changing production schedules, they have changed the structure of their energy usage towards lower-carbon production, and the higher the levy of carbon tax, the greater the effect of carbon reduction.
5. However, when the carbon tax is too high, the high cost of tax barriers will occur and the output need to meet with the final demand of each product, the carbon reduction effect is limited. In terms of cost, the levy of carbon tax can trigger the effect of CO_2 reduction.
6. When the carbon reduction cost is equal to the taxable cost, the carbon tax effect of CO_2 reduction will result. The profit and loss balance point is the best carbon tax level. If the carbon tax level is greater than the best tax rate at the Profit & Loss balance, there is an incentive force the industry to make a mandatory reduction. After levying carbon tax, the results show that when the cost of tax is more than the cost of reducing CO_2 emissions, wine industry will change the strategy to achieve CO_2 emission reduction, and get the tax effect.

This result also shows that the main source of the CO_2 emissions is fuel oil, followed by wastewater, process emissions and electricity. The CO_2 emissions improved significantly during the period of 2008 - 2016, the fuel oil, electricity, diesel, outsourcing and other emission had total decreased by 1.78%.The methods of the CO_2 reduction are:

1. Technical adjustments, such as packaging time in the off-season which is set from June to September, Centralizing manufacturing, reducing the times of boiler boot, blocking unnecessary pipelines, maintaining the steam pipe insulation facilities and strengthening the energy conservation advocacy. The most important thing is renewing the equipment.
2. This research also shows the intensity ratio indicator, in product density, service intensity and turnover intensity during the 2008-2016, there were a lower and flatter trend.

References

1. D. Babusiaux, A. Pierru, Appl. Energy. **84,** 828 (2007).
2. J.E. Aldy JE, Ley E, Parry IWH, Natl. Tax. J **61,** 493 (2008).
3. A.T.N. Moghaddam, C. Michelot, Eur. J. Oper. Rcs **197,** 999 (2009).
4. C.F. Lee, S. J. Lin, C. Lewis, Energy. Policy **36,** 722 (2008).
5. S. Soleille, Energ. Policy **34,** 1473 (2006).
6. S. Soimakallio, L. Saikku, Energy **38,** 3 (2012).
7. M.S. Chang, Clean Technol Envir **16,** 1001 (2014).
8. A. Omu, R. Choudhary, A. Boies, Energy. Policy **61,** 249 (2013).
9. Z. Wen, F. Meng, M Chen, J. Clean. Prod, **65,** 120 (2014).

Analysis of Energy Consumption of Indonesian Flat Glass Industry PT. X Based on Green Industry Standards

Reviana Revitasari[1] and Bambang Heru Susanto[1]

[1]*Department of Chemical Engineering, Faculty of Engineering, Universitas Indonesia, Depok 16424, Indonesia*

Abstract. The natural resource consumption is higher along with the increasing needs of people in various sectors, which affect the environment directly or indirectly. Especially for energy consumption, the supply and availability of fossil energy, as a non-renewable energy, are increasingly depleted and not guaranteed in the long run, coupled with the issue of emissions generated. The industrial sector as a driver of the national economy, including the flat glass industry as a case study, is an energy intensive sector. Almost 83% of its energy consumption used for operating the furnaces as the main production unit with temperature up to 1700 ^{0}C continuously about 15 years. The increasing of effectiveness and efficiency of energy consumption in the glass manufacturing process will affect significantly to sustainable production in PT. X and gives many other valuable impacts to the economic growth, environment, and society. So, this study analyzes about the energy consumption in PT. X based on Green Industry Standards, believed as a proper strategy, the benchmark of some standards or related regulations for energy consumption in several countries, and the opportunity of the green industry concept implementation in the glass manufacturing process. The methods were studied of literature, plant observation, interview, and data calculation manually and using spreadsheets. The results indicate that the Flat Glass Industry PT. X requires the improvement to increase the effectiveness and efficiency of energy consumption to get the sustainable production.

1 Introduction

The industrial sector as a driver of the national economy, including the flat glass industry as a case study, is an energy intensive sector. One of the objectives of industrial operation in Indonesia as stated in Law No.3/ 2014 on Industry is to realize an independent, competitive, and advanced industry. In addition, it is expected also the realization of a green industry. It is certainly in line with many global agreements in order to support greenhouse gas (GHG) mitigation, energy security promotion, and solution for various environmental issues and natural resource depletion nationally and globally. In order to support it, Indonesia government set the Green Industry Standard (GIS) to guiding, monitoring, and evaluating easily with measurable evaluation. Nowadays, it still voluntary program, but it will be a mandatory soon. GIS is a result of consensus from several stakeholders, that the representative of government, academia, industry players, and public figure. So that there will be adjustments within the agreed-upon standard, which will be different in other countries. So, it becomes one of the focus reviews in this paper.

Green industry, as stated in the Law of the Republic of Indonesia Number 3 2014 about Industry, is industries that in the production process, prioritize the efforts of efficiency and effectiveness of resource use (raw material, energy, and water) in a sustainable manner so as to harmonize the development of industry with the sustainability of environmental functions and can provide benefits for the community [1]. This paper focuses on Energy consumption in the glass manufacturing process as a significant point in the production process at Flat Glass Industry.

Flat glass industry PT. X produces flat or sheet glass with a production capacity of 612,500 tons per year. The process technology applied is a Float Process method which has 5 process units, that batch plant, melting furnace, tin bath, annealing Lehr, and cutting line. Raw material composition of glass in PT. X is Silica sand, soda ash, dolomite, feldspar, salt cake, limestone, sodium nitrate, cobalt oxide, nickel oxide, and cullet. Cullet is glass waste reused as raw material for glass manufacture. Its function is very important to reduce energy and raw material consumption with a total composition of 1 mix raw material consists of 25% of cullet and 75% of batch composition.

The mixture of feedstock is fed into the furnace and melted at about 1700 ^oC to form molten glass using a gas/ fuel burner.The furnace output temperature is about 1100 ^oC. Then the molten glass is fed to a surface of an 1100 ^oC liquid tin covered in liquid called tin bath. Temperature output of Tin bath is about 600 ^oC. Gradually, the glass is cooled from 600 to 100 ^oC, after that it is brought to the cutting area. In this area, automatic and manual defects are identified. Glass is cut in jumbo size or as per consumer demand. The edges of the glass bands are cut and recycled into the melting furnace as a cullet [2].

For a CO_2 emissions, about 75% of CO_2 emissions from float glass process are sourced from fossil fuels used to burn the furnace, while 25% comes from process emissions. The

CO_2 source of process emissions is not a function of efficiency, but a chemical process and therefore some emissions will remain, unless new processes are built. However, the use of recycled glass (lower use of raw materials) is increasing, resulting in a reduction in process emissions. There is also an indirect emissions from electricity use (melting and downstream activities). Emissions are not directly as a whole of the sheet glass sub-sector of about 16% of total CO_2 emissions per finished product [3, 4].

The purpose of this research is to get the overview and evaluation of energy consumption in Flat Glass Industry PT. X based on green industry standards and find the opportunity of decreasing energy consumption and implementation of the green industry concept there.

2 Experimental details

This research has begun by conducting a literature review of topics related to the research to understand the various theories that exist in this study, to understand all the issues related to the theme raised and obtain novelty aspects, as well as to determine methodology appropriate to the condition and availability of data in Indonesia, especially for Flat Glass Industry. After that, field observations and interviews have been made to obtain primary and secondary data that have been calculated manually and using spreadsheets. Energy distribution analysis made by using Sankey diagram software with goals as follows:

- Obtain a flow map of the energy distribution with clear and complete definition of each input and output in the whole production process.
- Know the units of processes with a large level of energy consumption and a strong influence on the level of energy consumption in the production process.
- Find out which operating units will be the main targets and have a significant impact on decision making.

The primary data are collected from measurements on energy measuring devices (flow meter, kWh meter) and plant observations and interviews. The secondary data are gotten from the energy balances (heat and electricity) and actual production data. The limits of the calculated heat and electricity energy consumption are the consumption of heat and electrical energy used for the production process (including utilities), but excluding those used for the office and domestic. The total energy consumption calculated is the sum of heat energy and electricity.

Then PT. X has been evaluated by Green Industry Standard of Flat Glass Industry to know the achievement of PT. X and formulated the recommendation.

The other limitations in this paper are:

a. The data used in this research is taken from Flat Glass Industry PT. X in an industrial area in Indonesia
b. This study focuses on energy consumption for production process of Float 2 (two) in PT. X.
c. The benchmark standards in this study were compared from the following standards or regulations:
   1) Green Industry Standard (GIS) from The Ministry of Industry, Indonesia.
   2) China Cleaner Production Standard of Flat Glass Industry: HJ/T361-2007.
   3) Intergovernmental Panel on Climate Change (IPCC) for Glass Industry
   4) Best Available Techniques (BAT) Reference Document for the Manufacture of Glass compiled by the European Commission.
   5) Final Report for a study on composition and drivers of energy prices and cost in Energy Intensive industries: The Case of The Flat Glass Industry by Centre for European Policy Studies (CEPS)
   6) United States Environment Protection Agency (EPA) and Department of Energy (DOE).

3 Results and discussion

3.1 Benchmark of standards

Especially in energy sector, there are many standards or regulations have been set in many other countries. The following figure displays the benchmark of some standards globally related to specific total energy consumption in Flat Glass Industry.

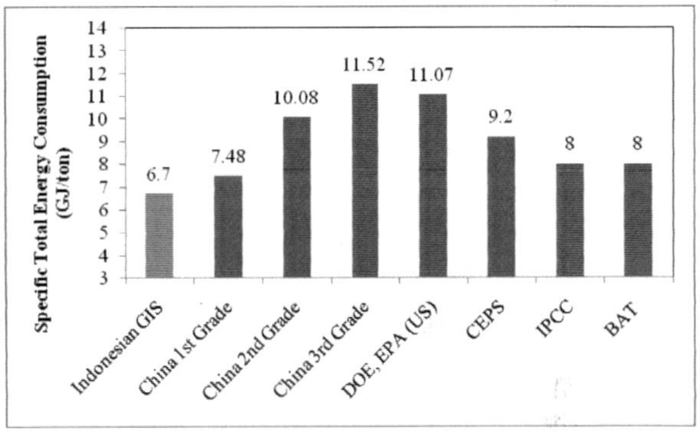

Figure 1. Benchmark of Maximum Specific Total Energy Consumption Standards [3-9].

Figure 1 shows that Indonesian Green Industry Standard is the most stringent standard related to maximum specific energy consumption. Following by China 1st grade standard, IPCC, BAT, China 2nd grade standard, DOE (US), and the highest is China 3rd standard. It indicates that the strong commitment of the Indonesian government, industry players, and academician to support the goals of green industry realization nationally and globally. So, in this paper, Indonesia GIS used as the standard to evaluate the Flat Glass Industry PT. X., that the total energy consumption per sheet of glass products (including cullet) in GIS is a maximum of 6.7 GJ/ton [7].

In Indonesia Green Industry Standard (GIS), the total energy consumption for production process is divided into 2 energy sources that Electricity and Heat Energy sourced from natural gas and fuel. The following table displays the limitation for each energy consumption in the flat glass industry.

Table 1. Standard limitation of specific energy consumption in flat glass industry in indonesian green industry standard (GIS).

Criteria	Standard Limitation	Unit
Specific Electricity Consumption	Max 0.7	GJ/ton
Specific Heat Energy Consumption	Max 6	GJ/ton
Specific Total Energy Consumption	Max 6.7	GJ/ton

So, the energy consumption in Flat Glass Industry PT. X analyzed and evaluated separately as Table 1, which has distinguished discussion based on its energy source.

3.2 Energy distribution and consumption

Flat glass industry PT. X has 3 Float Process Units. Each float produces different glass type and capacity based on process technology, sources of energy, and glass product demand. So that, to get a brief and comprehensive overview and evaluation, conducted the analysis for each float separately. This paper focuses on energy consumption of Float 2.

As described before that flat glass industry is one of the energy intensive sectors. The dominant energy used in the flat glass industry PT. X is natural gas and petroleum. The additional energy is electrical energy. Natural gas is the main fuel for glass production, followed by petroleum products. Both fuels are exchanged in the melting process. More than three quarters of the energy used in the float process comes from glass melting. Furnaces as its main operating unit that operates continuously, and uses up to 83% of the total energy required for production processes in the industry and operating continuously up to 15 years. The most energy intensive stage of the glass sheet production process is the activity of the furnace, when heating uses natural gas. Figure 2 displays the energy distribution of Float 2 in Flat Glass Industry PT. X. The energy source of Float 2 are Electricity obtained from the State Electricity Company, natural gas, Diesel, and Residue.

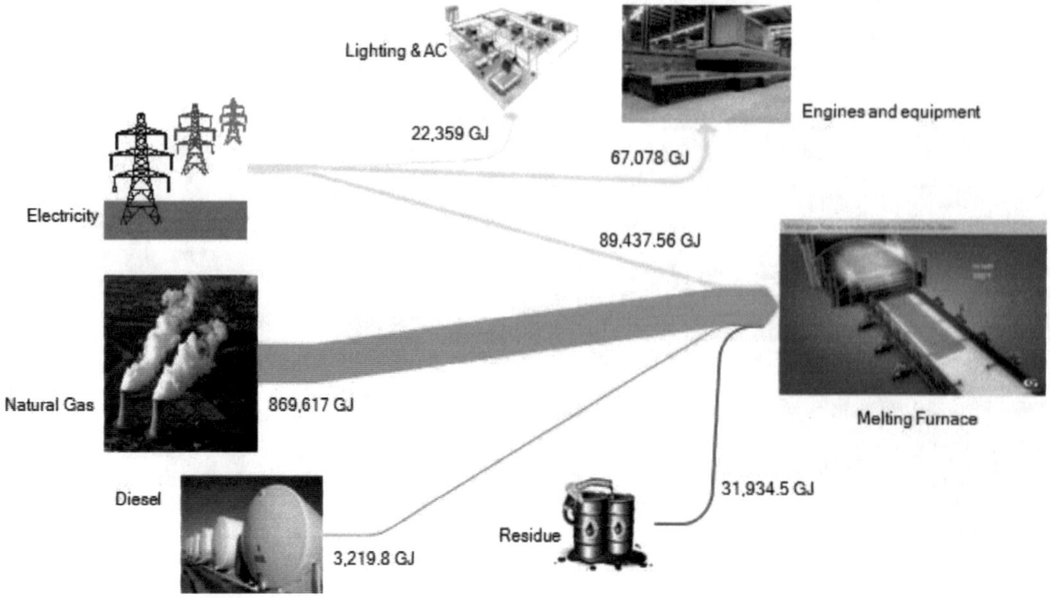

Figure 2. Energy Distribution of Float 2 in Flat Glass Industry PT. X in 2016

From the above figure, it is proven that the highest energy consumption is in the hot end process which melting furnace as the main process unit. So, the improvement recommendation mostly related to the melting process, which is also affected by the previous process.

3.2.1 Specific heat energy consumption

The specific heat energy consumption in glass manufacturing is heat energy consumption per sheet of glass products (including cullet). The heat energy consumption data is a total of natural gas, diesel, and residue consumption. The following figure shows the specific heat energy consumption of Float 2 in PT. X.

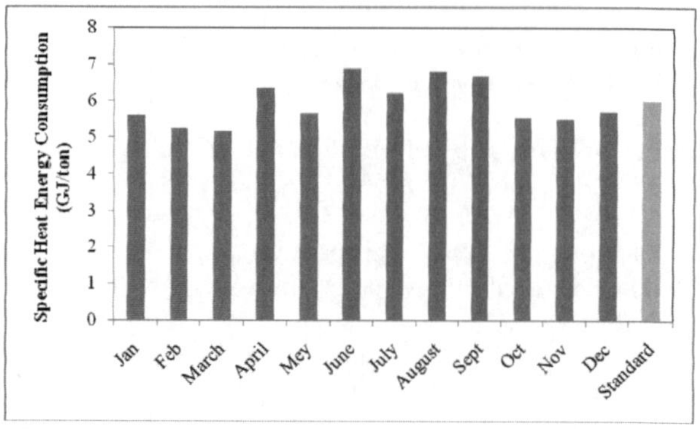

Figure 3. The Specific Heat Energy Consumption of Float 2 in Flat Glass Industry PT. X by month in 2016.

Figure 3 shows energy consumption of PT. X for 7 months among 2016 under the Indonesian Green Industry Standard. But, there are 5 months which is higher than the limitation standard. As for the standard consumption of heat energy per sheet of glass products (including cullet) is a maximum of 6 GJ/ton [7]. Meanwhile the specific heat energy consumption of PT. X in that 5 months are 6.36, 6.88, 6.22, 6.82, 6.69 GJ/ton. The average specific heat energy consumption in 2016 is 5.95 GJ/ton.

3.2.2 Specific electrical energy consumption

The specific electrical energy consumption is consumption of electrical energy per sheet of glass products (including cullet). The figure 3 displays the specific electrical energy consumption in float 2 of Flat Glass Industry PT. X by month in 2016.

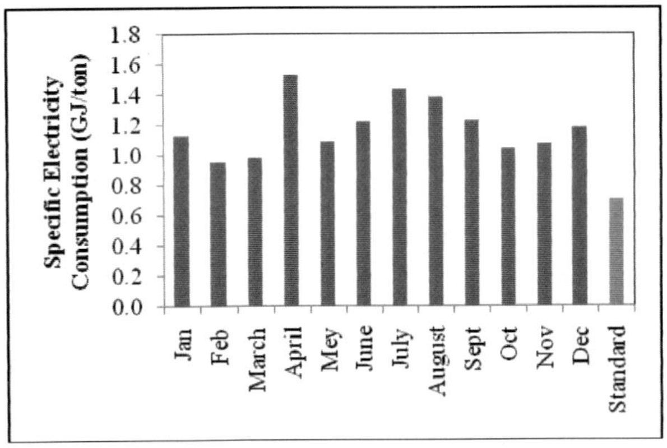

Figure 4. The Specific Electricity Consumption of Float 2 in Flat Glass Industry PT. X in 2016.

For the consumption of electrical energy per sheet of glass products (including cullet) according to the Green Industry standard is a maximum of 0.7 GJ/ton [7]. Figure 4 shows that electrical energy consumption of Float 2 PT. X is higher than limitation standard in each month. The lowest electricity consumption is 0.95 GJ/ton in February and the highest is 1.53 GJ/ton in April 2016. The average specific electricity consumption in 2016 is 1.19 GJ/ton.

3.2.3 Specific total energy consumption

The total energy consumption is the sum of heat and electrical energy consumption. Then the specific total energy consumption is the total energy consumption per sheet glass product (including cullet). The figure below displays it compare to limitation standard in Indonesian GIS.

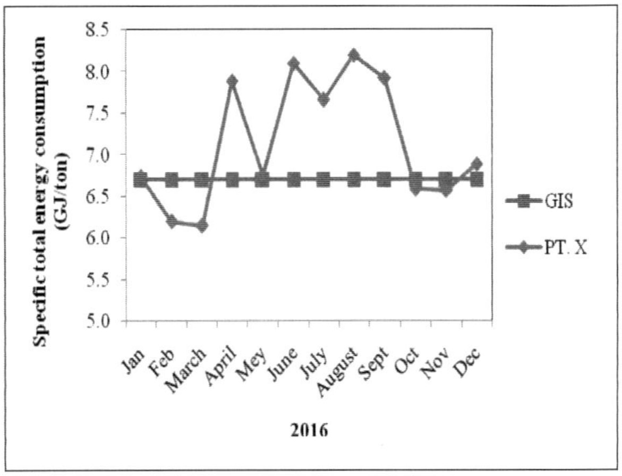

Figure 4. The Specific Total Energy Consumption of Float 2 in Flat Glass Industry PT. X in 2016.

The maximum consumption standard of total energy consumption per sheet of glass products (including cullet) in GIS is 6.7 GJ/ton [7]. The result shows that almost every month, its specific total energy consumption is higher than the GIS standard. The average specific total energy consumption in PT. X in 2016 is 7.14 GJ/ton.

Based on plant study, survey, interview, and data calculation, found several things that cause the amount of energy consumption in float 2, such as:

a. Changes in glass product demand while operation unit operate continuously
b. Heat loss in the furnace wall
c. No batch and or cullet preheating process before melting process
d. Changes in cullet usage

The use of cullet in a glass furnace can significantly reduce the energy consumption and its use is generally applicable to all types of furnaces. Cullet has a lower melting energy requirement than the constituent raw materials because endothermic chemical reactions associated with glass formation have been completed and its mass is approximately 20 % lower than the equivalent batch materials. Therefore, increasing the cullet level in the batch has the potential to save energy and, as a general rule, every 10 % of extra cullet results in a 2.5 – 3.0 % reduction in furnace energy consumption. The use of cullet generally results in significant cost savings as a result of the reduction in both energy and raw material requirements [2]. The use of cullet in PT. X depends on the avaibility of internal cullet or cullet in house that glass shards originating from the factory waste itself so as to have a glass composition advantage similar to that produced by the factory. PT. X also use of external cullet called cullet foreign. That cullet from outside the factory or from waste collecting stalls are usually different in composition than factory-produced materials. Their storage must be separated due to different composition. The cullet should also be free of contamination such as oil, ash, metal, packaging materials and others as it will affect the quality of the resulting glass [2]. So, the amount of cullet is limited by the availability of the right quality cullet and proper chemical compatibility.

4 Summary

Flat glass industry is an energy intensive sector. Almost 83% of its energy consumption used for operating the furnaces as the main production unit with temperature up to 1700 ^{0}C continuously about 15 years. Besides, its primary energy source is fossil energy. Green industry concept believed as the most proper strategy in encouraging manufacturing industries towards using sustainable patterns of production and economic growth. Green Industry Standard (GIS) was set to guiding, monitoring, and evaluating industry easily with measureable evaluation. It is a consensus result from several stakeholders, that the representation of government, academician, industry players, and public figure. So that there have been adjustment within the agreed-upon standard. In energy sector, limitation standard of Indonesian GIS is the most stringent standard related to maximum specific energy consumption among China Clean production standard, IPCC, BAT, regulation of DOE and EPA (US). It indicates that the strong commitment of Indonesian government, industry players, and academician to support the goals of green industry realization nationally and globally. So, in this paper, Indonesia GIS used as the standard to evaluate the Flat Glass Industry PT. X.

The results showed that the Flat Glass Industry PT. X needs the improvement programs to increase the effectiveness and efficiency of energy consumption, heat as well as electricity. Based on plant study, survey, interview, and data calculation, found several things that cause the amount of energy consumption in float 2, such as:

a. Changes in glass product demand while operation unit operate continuously
b. Heat loss in the furnace wall
c. No batch and or cullet preheating process before melting process
d. Changes in cullet usage

The valuable opportunity in PT. X which is expected increase the effectiveness and efficiency of sustainable production is by increasing the effectiveness and efficiency of the energy. It will have a very significant effect in the decrease the energy consumption, which leads to a reduction in the cost of glass production and GHG emissions. Some alternatives are:

a. Stabilize the Batch and Cullet ratio (B/C ratio)
b. Apply the batch and or cullet preheating process
c. Repair or add the addition insulation in the melting furnace

So by increasing the efficiency of the furnace or melting process will give a very significant effect in the decrease in energy consumption, which leads to a reduction in glass production costs and GHG emissions caused.

The last, the implementation of green industry standards is highly feasible and rewarding, proven with the achievements that have been achieved by PT. X as case study. It can help the industry and government to conduct monitoring and evaluation of the industry as a whole and comprehensive way. So it needs an active cooperation between industry players, academicians, the government and society in realizing a green industry based on the best standards. This study had to be continued to be developed for the advancement of science and expediency for other sectors.

Acknowledgement

The authors wish to deep sense of gratitude towards Universitas Indonesia, LPDP (Indonesia Endowment Fund for Education), and PT. X.

References

1. Law of the Republic of Indonesia Number 3 2014 about Industry
2. PT. X, *Report Document of Flat Glass Industry,* (Bekasi, Indonesia, 2017)
3. S. B. Maria, G. M. Marcos, S. A. Querol, R. Serge, D. S. Luis, *JRC Reference Report: Best Available Techniques (BAT) Reference Document for the Manufacture of Glass*. (Publications Office of the European Union, Luxembourg, 2013)
4. C. Egenhofer and L. Schrefler, *Final Report for a Study on Composition and Drivers of Energy Prices and Cost in Energy Intensive Industries: the Case of the Flat Glass Industry* (Centre for European Policy Studies, Brussels, 2014)
5. Intergovernmental Panel on Climate Change (IPCC) for Glass Industry
6. Ministry of Environmental Protection, the National Environmental Protection Standard of the Republic of China, *Clean Production Standard of Flat Glass Industry. HJ/T361-2007*(2007)
7. The Ministry on Industry, Republic of Indonesia, *Green Industry Standard, Flat Glass Industry, SIH 23111.1:2016* (2016)
8. United States Department of Energy (DOE), *Specific Energy Consumption US Flat Glass* (2002)
9. United States Environment Protection Agency (EPA), *Specific Energy Consumption US Flat Glass* (2008)

Electrostatic Force between two Dielectric Particles in Electrorheological Fluids: beyond Spherical Particles

Hongzhe TANG[1], Li GAN[1] and Wei AN[2]

[1]*Laboratoire International Associé, Ecole Centrale de Pékin, Beihang University, Beijing 100191, China*
[2]*Laboratory of Mathematics and Physics, Ecole Centrale de Pékin, Beihang University, Beijing 100191, China*

Abstract. In an effort to increase the shear yield stress of dielectric electrorheological fluids, we focus on the electrostatic force of different forms of particles in a dielectric polarization model. By solving Laplace's equation and applying the multiple image method and the finite element method, the analytical and numerical solutions of the electrostatic force of a two-sphere structure have been studied. The results suggest that when the dielectric mismatch factor is large and when the positions of the two spheres are nearly in contact with each other, most of the analytical solutions either over- or underestimate the force. Additionally, the structure of particles beyond the spherical form is considered. Three example cases are studied to shed light on how different geometries of particles may affect the electrostatic force, thereby influencing the shear yield stress of the fluid.

Nomenclature

a, b	radius of two spheres, m
d	distance between the centers of two spheres, m
$\mathbf{D}$	electric displacement field, $\mathrm{C \cdot m^{-2}}$
$\mathbf{E}$	electric field, $\mathrm{V \cdot m^{-1}}$
$\mathbf{F}$	force, N
h	Planck constant, $\approx 6.626 \times 10^{-34}$ $\mathrm{J \cdot s}$
k_B	Boltzmann constant, $\approx 1.381 \times 10^{-23}$ $\mathrm{J \cdot K^{-1}}$
$\mathbf{p}$	electric dipole moment, $\mathrm{C \cdot m}$
$\mathbf{r}$	displacement, m
$\hat{\mathbf{r}}, \hat{\boldsymbol{\theta}}$	local orthogonal unit vectors in the directions of increase r, θ in spherical coordinate system
S	configuration entropy, $\mathrm{J \cdot K^{-1}}$
T	thermodynamic temperature, K
V	volume, $\mathrm{m^{-3}}$
w	electric energy density, $\mathrm{J \cdot m^{-3}}$
W	electric potential energy, J
Greek symbols	
β	Clausius-Mossotti factor, $(\kappa - 1)/(\kappa + 2)$
γ	ratio of forces between the DID and PD
ϵ	relative permittivity
ϵ_0	vacuum permittivity, $\approx 8.854 \times 10^{-12}$ $\mathrm{F \cdot m^{-1}}$
κ	dielectric mismatch factor, ϵ_p / ϵ_f
λ	wavelength m ; scale factor
ρ	volume charge density, $\mathrm{C \cdot m^{-3}}$
σ	surface charge density, $\mathrm{C \cdot m^{-2}}$
τ	spacing factor, *i.e.*, d / a
ϕ	electric potential, V
Superscripts	
(L)	longitudinal electric field, *i.e.*, the applied field that is parallel to the line joining the centres of the spheres
(T)	transverse electric field, *i.e.*, the applied field that is perpendicular to the line joining the centres of the spheres
Subscripts	
f	fluid
p	particle

1 Introduction

Electrorheological (ER) fluids are regarded as smart fluids, whose resistance to flow (from a liquid to a solid gel, and back) can be quickly (with response times on the order of milliseconds [1]) and dramatically altered by an applied electric field. Several applications have been proposed, including for clutches, brakes, shock absorbers, hydraulic valves, polishing, and artificial muscles [2-9]. In this work, we consider dielectric electrorheological (DER) fluids, which are made of micrometer-sized dielectric particles in an electrically insulating liquid. Compared to that of magnetorheological (MR) fluids, the yield stress of DER fluids is much lower. The shear yield stress of a general DER material is typically of several kPa when subjected to an electric field of 1-5 $\mathrm{V \cdot mm^{-1}}$

[10], while the typical shear yield stress of MR fluid could vary from 2-3 kPa with no magnetic field to 50-100 kPa with an applied magnetic field of 240 $\mathrm{A\cdot m^{-1}}$ [11]. The DER fluid has therefore found limited use in the mechanical industry.

Our aim is to increase its yield stress, which, at the micro level, means to increase the interparticle force (mainly the electrostatic force, and other forces are discussed in Refs. [12-14]) according to the dielectric polarization model (DPM), where the electrical response is governed by linear electrostatics. Recently, dielectric effects have gained increased attention in other fields. For instance, E. Luijten *et al.* investigated polarization effects in charge-stabilized dielectric colloidal suspensions. They introduced an efficient technique to resolve polarization charges in dynamical dielectric geometries and demonstrated that dielectric effects qualitatively alter the predicted self-assembled structures [15, 16].

In this work, we ignore temperature effects and entropic effects, *i.e.*, the first and the third term in the expression of the free energy W [17, 18]

$$W = \frac{3}{2}Nk_{\mathrm{B}}T - \frac{1}{2}\int_{\mathrm{particles}}(\mathbf{ED} - \mathbf{E}_0\mathbf{D}_0)\,d^3\mathbf{r} - TS, \quad (1.1)$$

where N is the total number of dielectric particles in the DER fluid; k_{B} is the Boltzmann constant; T is the thermodynamic temperature, which is usually the room temperature; and $\mathbf{E}, \mathbf{D}(\mathbf{E}_0, \mathbf{D}_0)$ are, respectively, the local electric field and electric displacement with and without the dielectric particles. S is the configuration entropy, which can be well-approximated by

$$S = Nk_{\mathrm{B}}\ln\left(\frac{\Omega}{N\lambda_{\mathrm{th}}^3}\right), \quad (1.2)$$

where Ω is the volume available to the many-particle system and the thermal de Broglie wavelength is $\lambda_{\mathrm{th}} = h/\sqrt{3mk_{\mathrm{B}}T}$. h is the Planck constant. Consider a spherical particle at 300 K with a radius of 5 μm and a density of $4\ \mathrm{g\cdot cm^{-3}}$. The thermal wavelength is $\lambda_{\mathrm{th}} \approx 4.11\times10^{-18}$ m, which is very small compared to the size of the particle. Generally, the particle diameters range from 0.1 to 100 μm [13]. In fact, the electrostatic energy usually dominates the thermal and entropic effects [19], *i.e.*, the electrostatic energy of a particle is much greater than $k_{\mathrm{B}}T$ and $k_BT\ln\left(\Omega/\left(N\lambda_{\mathrm{th}}^3\right)\right)$.

In the dipole DPM, the interparticle force depends on the electric dipole moment. Thus, it has inspired us to focus on the hierarchical structure of the particles (stick, lamella, needle-like, etc.) assembled at different scales. Charges polarized in an electric field tend to concentrate at certain positions of the particles surface so that the charges can be used efficiently to increase the shear yield stress. By optimizing the hierarchical structure combination, a high performance electrorheological fluid can be efficiently fabricated. To the best of our knowledge, work of such nature has not yet been performed. Various investigations, such as of particle patterns, arrangement structures, surface/core dielectric properties and interface structures, could be conducted to characterize the fluids. With a better characterization of electrorheological fluids, they may find wider applications in various domains, such as aviation, noise control, and automobile engineering [20].

In previous papers, the separation of variables and multiple image methods are used to determine the analytical formula of the electrostatic force between two spherical particles in the DPM [21-23]. In this paper, by proposing to investigate the problem via the finite element method (FEM) approach, we are no longer confined by the regular shape of the particles and the other parameters that we study.

In the first part, we considered the rudimentary case of the model, *i.e.*, the point-dipole (PD) model. Then, in the following section, we consider models beyond the PD case. Next, we proceed to consider three special cases beyond the basic form as a demonstration, which provides several ideas to increase the shear yield stress in the experiment.

2 Solution in the point-dipole model

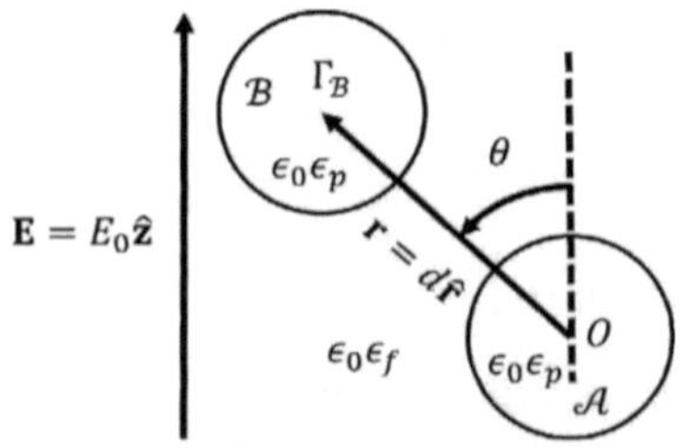

Figure 1. Model of two dielectric spheres in DER fluids.

Based on Sect. 4.4 of Ref. [24], we derive the formula of the electrostatic force in the point-dipole model in a completely rigorous way. As shown in Figure 1, we consider two uncharged dielectric spheres $\mathcal{A}$ and $\mathcal{B}$, of radius a, having identical dielectric constants $\epsilon_0\epsilon_p$, fixed in place in a liquid, with a dielectric constant $\epsilon_0\epsilon_f$, far from the electrodes so that it appears to be in a uniform electric field $\mathbf{E} = E_0\hat{\mathbf{z}}$. We assume $\mathcal{A}$ and $\mathcal{B}$ as two electric dipoles: $\mathcal{A}$ is at the origin, with a dipole moment $\mathbf{p}_1$; $\mathcal{B}$, with a dipole moment $\mathbf{p}_2$, has a displacement $\mathbf{r} = d\hat{\mathbf{r}}$ from $\mathcal{A}$. Additionally, we assume the two particles are independent of each other. We note $\mathbf{E}_1$ the electric field generated by $\mathcal{A}$. The force $\mathbf{F}$ that $\mathcal{A}$ exerts on $\mathcal{B}$ can be written as

$$\mathbf{F}_{\mathcal{A}\to\mathcal{B}} = -\nabla W = \nabla(\mathbf{p}_2\cdot\mathbf{E}_1). \quad (2.1)$$

$\mathbf{p}_2$ and $\mathbf{E}_1$ may be calculated by studying the electrostatic problem of particle $\mathcal{A}$. Since there are no free charges ($\rho = 0$) in the exterior or interior of the sphere, the potential around $\mathcal{A}$, noted ϕ , satisfies Laplace's equation

$$\nabla^2\phi = -\frac{\rho}{\epsilon} = 0, \quad (2.2)$$

with the boundary conditions

$$\begin{cases} \phi_{\text{in}}\big|_{r\to 0} < \infty, \\ \phi_{\text{out}}\big|_{r\to\infty} = -E_0 r\cos\theta, \\ \text{Tangentiel } \mathbf{E}: \ -\frac{1}{a}\frac{\partial \phi_{\text{in}}}{\partial\theta}\bigg|_{r=a} = -\frac{1}{a}\frac{\partial \phi_{\text{out}}}{\partial\theta}\bigg|_{r=a}, \\ \text{Normal } \mathbf{D}: \ -\epsilon_p \frac{\partial \phi_{\text{in}}}{\partial r}\bigg|_{r=a} = -\epsilon_f \frac{\partial \phi_{\text{out}}}{\partial r}\bigg|_{r=a}. \end{cases} \quad (2.3)$$

By separation of variables and azimuthal symmetry, the general solution, called the harmonic function, can be written in the form of Legendre polynomials, and the solution of this problem can be found using the method of undetermined coefficients. We note the dielectric mismatch factor $\kappa = \epsilon_p / \epsilon_f$. The potential is therefore

$$\begin{cases} \phi_{\text{in}} = -\left(\frac{3}{\kappa+2}\right)E_0 r\cos\theta, \\ \phi_{\text{out}} = -E_0 r\cos\theta + \beta E_0 \frac{a^3}{d^2}\cos\theta, \end{cases} \quad (2.4)$$

where the Clausius-Mossotti factor is $\beta = \frac{\kappa-1}{\kappa+2}$.

Then, we determine the relation between the potential and the dipole moment

$$\phi(\mathbf{r}) = \sum_i \frac{q_i}{4\pi\epsilon_0\epsilon_f}\frac{1}{|\mathbf{r}-\mathbf{r}_i|} \approx \sum_i \frac{q_i}{4\pi\epsilon_0\epsilon_f}\left(\frac{1}{d} - \mathbf{r}_i\cdot\nabla\left(\frac{1}{d}\right)\right) = \frac{1}{4\pi\epsilon_0\epsilon_f}\frac{\mathbf{p}\cdot\hat{\mathbf{r}}}{d^2}. \quad (2.5)$$

Hence,

$$\phi_{\text{out}} = -E_0 r\cos\theta + \frac{1}{4\pi\epsilon_0\epsilon_f}\frac{\mathbf{p}_2\cdot\hat{\mathbf{r}}}{d^2}, \quad (2.6)$$

and

$$\mathbf{p}_2 = 4\pi\epsilon_0\epsilon_f a^3 \beta\mathbf{E} = \mathbf{p}_1 := \mathbf{p} = p\hat{\mathbf{z}}. \quad (2.7)$$

After determining $\mathbf{p}_2$, $\mathbf{E}_1$ may be calculated using

$$\mathbf{E}_1 = -\nabla(\phi_{\text{out}} + E_0 r\cos\theta) = \frac{1}{4\pi\epsilon_0\epsilon_f}\left(\frac{3(\mathbf{p}\cdot\hat{\mathbf{r}})\hat{\mathbf{r}} - \mathbf{p}}{d^3}\right) - \frac{1}{3\epsilon_0\epsilon_f}\mathbf{p}\delta^3(\mathbf{r}), \quad (2.8)$$

where $\delta^3(\cdot)$ is the 3-dimensional Dirac delta function. Since we study the case outside of $\mathcal{A}$, *i.e.*, $\mathbf{r} \neq 0$, we have

$$\mathbf{E}_1 = \frac{1}{4\pi\epsilon_0\epsilon_f}\left(\frac{3(\mathbf{p}\cdot\hat{\mathbf{r}})\hat{\mathbf{r}} - \mathbf{p}}{d^3}\right). \quad (2.9)$$

Eventually, we obtain the expression of the force $\mathbf{F}$ that $\mathcal{A}$ exerts on $\mathcal{B}$ and the dimensionless force $\tilde{\mathbf{F}}_{\mathcal{A}\to\mathcal{B}}$.

$$\mathbf{F}_{\mathcal{A}\to\mathcal{B}} = \frac{3p^2}{4\pi\epsilon_0\epsilon_f d^4}\left[(1-3\cos^2\theta)\hat{\mathbf{r}} - \sin 2\theta\hat{\boldsymbol{\theta}}\right],$$

$$\tilde{\mathbf{F}}_{\mathcal{A}\to\mathcal{B}} := \frac{\mathbf{F}_{\mathcal{A}\to\mathcal{B}}}{4\pi\epsilon_0\epsilon_f (E_0 a)^2} = 3\beta^2\tau^{-4}\left[(1-3\cos^2\theta)\hat{\mathbf{r}} - \sin 2\theta\hat{\boldsymbol{\theta}}\right], \quad (2.10)$$

where $\tau = d/a$.

Specially, (i) when $\theta = 0$, the electric field is parallel to the line joining the centers of the spheres, and $\tilde{\mathbf{F}}^{(L)}_{\mathcal{A}\to\mathcal{B}} = -6\beta^2\tau^{-4}$. The negative value denotes attraction; (ii) when $\theta = \frac{\pi}{2}$, the electric field is perpendicular to the line joining the centers of the spheres, and $\tilde{\mathbf{F}}^{(T)}_{\mathcal{A}\to\mathcal{B}} = 3\beta^2\tau^{-4}$. The positive value represents repulsion; (iii) when two spheres are in direct contact with each other, $\tilde{\mathbf{F}}_{\mathcal{A}\to\mathcal{B}} = \frac{3}{16}\beta^2\left[(1-3\cos^2\theta)\hat{\mathbf{r}} - \sin 2\theta\hat{\boldsymbol{\theta}}\right]$, and the two cases above become $-\frac{3}{8}\beta^2$ and $\frac{3}{16}\beta^2$, respectively, which are both limited values.

Additionally, according to the DPM, assuming the pairwise additivity and only nearest neighbor interactions, the yield stress varies as E_0^b, where E_0 is the scalar quantity of the external uniform electric field, and $b \approx 2$ is commonly observed for low to moderate field strengths. At large field strengths, b can decrease below 2 over a range in $\kappa = \epsilon_p / \epsilon_f$ (3-500) [13], which can be explained by the nonlinear conduction [25].

3 Solution beyond the point-dipole model

Beyond the dipole approximation, to solve the problem of two dielectric spheres completely, the exact analytical solution can be calculated in bispherical coordinates, which is a coordinate system in which the separation of variables is possible [21].

A much easier method is to apply the multiple image method to attain the multipole-induced-dipole (MID) model. The nomenclature is similar to that of the point-dipole model. In this case, however, we consider $\mathcal{A}$ and $\mathcal{B}$ with different radii a and b and different Clausius-Mossotti factors β_a and β_b, respectively. Note d, the distance between the centers of two spheres. Additionally, we no longer assume that the two particles are independent of each other. In Ref. [23], Siu *et al.* give the expression of the dipole moment of sphere $\mathcal{A}$ for two special cases: a transverse electric field (T) and a longitudinal electric field (L). Here, we introduce a feature parameter

$$\lambda = \begin{cases} -1 & \text{, for } (T) \text{ case,} \\ 2 & \text{, for } (L) \text{ case.} \end{cases} \quad (3.1)$$

Therefore, the expression of the dipole moment can be written in a unified form

$$p_{\mathcal{A}}^{(T)\text{ or }(L)} = (\sinh\alpha)^3 \sum_{n=1}^{\infty}\left\{\frac{p_{\mathcal{A}0} b^3 (\lambda\beta_a)^{n-1}(\lambda\beta_b)^{n-1}}{[b\sinh n\alpha + a\sinh(n-1)\alpha]^3} + \frac{p_{\mathcal{B}0} a^3 (\lambda\beta_a)^{n}(\lambda\beta_b)^{n-1}}{(d\sinh n\alpha)^3}\right\}, \quad (3.2)$$

where $p_{\mathcal{A}0}$ and $p_{\mathcal{B}0}$ are scalar quantities of the dipole moments in the point-dipole model

$$p_{\mathcal{A}0} = 4\pi\epsilon_0\epsilon_f\beta_a E_0 a^3, \ p_{\mathcal{B}0} = 4\pi\epsilon_0\epsilon_f\beta_b E_0 b^3. \quad (3.3)$$

The hyperbolic angle parameter α satisfies

$$\cosh\alpha = \frac{d^2 - a^2 - b^2}{2ab}. \quad (3.4)$$

The force between the spheres is given by

$$F^{(T)} = \frac{E_0}{2}\frac{\partial}{\partial d}\left(p_{\mathcal{A}}^{(T)} + p_{\mathcal{B}}^{(T)}\right), F^{(L)} = \frac{E_0}{2}\frac{\partial}{\partial d}\left(p_{\mathcal{A}}^{(L)} + p_{\mathcal{B}}^{(L)}\right). \quad (3.5)$$

The direction is along the line joining the centers of the spheres. The positive value represents repulsion, and the negative value represents attraction.

When n=1, it is the point-dipole case

$$F_{\mathrm{PD}}^{(T)\text{ or }(L)} = -12\pi\epsilon_0\epsilon_f\lambda\left(\frac{a^3}{d^2}\beta_a E_0\right)^2,$$
$$\tilde{F}_{\mathrm{PD}}^{(T)\text{ or }(L)} := \frac{F_{\mathrm{PD}}^{(T)\text{ or }(L)}}{4\pi\epsilon_0\epsilon_f\left(E_0 a\right)^2} = -3\lambda\beta_a^2\left(\frac{a}{d}\right)^4, \tag{3.6}$$

which is the same as Eq. (2.10) with $\theta = 0$ or $\frac{\pi}{2}$.

Considering the n=1 and n=2 terms, we obtain the expression in the dipole-induced-dipole (DID) model. The ratio γ of the forces between the DID and the PD models can be written as

$$\gamma^{(T)\text{ or }(L)} := \frac{F_{\mathrm{DID}}^{(L)}}{F_{\mathrm{PD}}^{(L)}} \text{ or } \frac{F_{\mathrm{DID}}^{(T)}}{F_{\mathrm{PD}}^{(T)}}$$
$$= 1 + \frac{\lambda\beta_a a^3 d^5}{\left(d^2-b^2\right)^4} + \frac{\lambda\beta_b b^3 d^5}{\left(d^2-a^2\right)^4} + \frac{\lambda^2\beta_a\beta_b a^3 b^3\left(3d^2-a^2-b^2\right)}{\left(d^2-a^2-b^2\right)^4}. \tag{3.7}$$

More terms can be considered, but the calculation becomes much more complex. The multipole-induced-dipole model includes all the terms. After extensive calculations, we obtain the expression of the electrostatic force in the MID model

$$\tilde{F}_{\mathrm{MID}}^{(T)\text{ or }(L)} := \frac{F_{\mathrm{MID}}^{(T)\text{ or }(L)}}{4\pi\epsilon_0\epsilon_f abE_0^2}$$
$$= \frac{3}{2}(ab)^2\left\{G\sum_{n=1}^{\infty}\left[\left(\lambda^2\beta_a\beta_b\right)^n\left(A+B\right)\right]\right. \tag{3.8}$$
$$\left.-\sinh^3\alpha\sum_{n=1}^{\infty}\left[\left(\lambda^2\beta_a\beta_b\right)^n\left(C+D\right)\right]\right\},$$

where

$$A = -\frac{2}{\left(d\sinh n\alpha\right)^3}, B = \sum_{\substack{i\neq j\\(i,j)\in\{a,b\}^2}}\frac{1}{\lambda^2\beta_i\left(i\sinh n\alpha + j\sinh(n-1)\alpha\right)^3},$$

$$C = \frac{2}{\lambda d^2\sinh^3 n\alpha}\left(\frac{1}{d^2} + \frac{n\cosh n\alpha}{ab\sin\alpha\sinh n\alpha}\right),$$

$$D = \sum_{\substack{i\neq j\\(i,j)\in\{a,b\}^2}}\frac{d\left(in\cosh n\alpha + j(n-1)\cosh(n-1)\alpha\right)}{\lambda^2\beta_i ab\sinh\alpha\left(i\sinh n\alpha + j\sinh(n-1)\alpha\right)^4},$$

$$G = \frac{d}{2(ab)^2}\cosh\alpha\sqrt{\prod_{(i,j)\in\{\pm1\}^2}(d+ia+jb)}.$$

Inspired by the PD model, the feature parameter λ could be extended as

$$\lambda = 3\cos^2\theta - 1, \tag{3.9}$$

and $\tilde{F}_{MID}$ can be generalized to the projection of the electrostatic force between two spheres in the direction of $-\hat{\mathbf{r}}$. Several limitations of Eq. (3.8) warrant comment. As it is an approximation in the MID model, more details will be discussed in the following part.

4 Preliminary validation works for the solution beyond the spherical form

So far, by confining ourselves to the hypothesis of an ideal spherical particle, there is a difference between the theoretical and experimental results: what about other forms of particle $\mathcal{A}$? By varying the boundary $\partial\Omega_A$, the case is transferred to a Dirichlet problem for Laplace's equation, whose analytical solution is difficult to calculate by adding correction terms as in the case of a two-sphere problem. An alternative approach is to use a numerical solution; the finite element method (FEM) is investigated for its feasibility in approximating the solution for problems of such nature.

Many researchers have considered cases beyond a typical two-sphere model. For instance, Wen *et al.* considered the coated sphere structure, and shown that the use of an effective dielectric constant concept based on the first-principles method can yield quantitative predictions [19]. Tao *et al.* adopted an infinite chain of dielectric particles [22], while Ahn *et al.* investigated n-sphere clusters [26]. J. G. Cao *et al.* considered the microstructure evolution from a random structure that consists of spherical particles to chains and then to stable lamellar patterns [27]. However, to the best of the authors' knowledge, there are few studies that consider models beyond that of spherical particles. Hence, before studying the non-spherical case, we start from the abovementioned spherical model, the simplest case, to verify its numerical solution accuracy when compared with the analytical solution.

Consider a pair of dielectric spheres $\mathcal{A}$ and $\mathcal{B}$ of identical radii a and dielectric constants $\epsilon_0\epsilon_p$ being embedded in an electrically insulating liquid with dielectric constant $\epsilon_0\epsilon_f$ in response to an applied electric field, $\mathbf{E} = E_0\hat{\mathbf{z}}$, which is parallel to the line joining the centers of the spheres. We note the spacing factor $\tau = d/a$. Eqs. (3.7) and (3.8) can be simplified as

$$F_{\mathrm{DID}}^{(L)} = \gamma F_{\mathrm{PD}}^{(L)} = \left[1 + \frac{4\beta\tau^5}{\left(\tau^2-1\right)^4} + \frac{4\beta^2\left(3\tau^2-2\right)}{\left(\tau^2-2\right)^4}\right]F_{\mathrm{PD}}^{(L)},$$
$$\tilde{F}_{\mathrm{DID}}^{(L)} := \frac{F_{\mathrm{DID}}^{(L)}}{4\epsilon_0\epsilon_f\left(E_0 a\right)^2}. \tag{4.1}$$

$$F_{\mathrm{MID}}^{(L)} = K\left[G'\sum_{n=1}^{\infty}(A'+B') - 4(\sinh\alpha)^3\sum_{n=1}^{\infty}(C'+D')\right] := K\sum_{n=1}^{\infty}F_n^{(L)},$$
$$\tilde{F}_{\mathrm{MID}}^{(L)} := \frac{F_{\mathrm{MID}}^{(L)}}{4\epsilon_0\epsilon_f\left(E_0 a\right)^2}, \tag{4.2}$$

where

$$A' = \frac{(2\beta)^{2n-1}}{\left(\sinh n\alpha + \sinh(n-1)\alpha\right)^3}, B' = \frac{(2\beta)^{2n}}{\left(\tau\sinh n\alpha\right)^3},$$

$$C' = \frac{(2\beta)^{2n-1}}{\sqrt{\tau^2-4}}\frac{n\cosh n\alpha + (n-1)\cosh(n-1)\alpha}{\left(\sinh n\alpha + \sinh(n-1)\alpha\right)^4},$$

$$D' = \frac{(2\beta)^{2n}}{\left(\tau\sinh n\alpha\right)^3}\left(\frac{n\coth n\alpha}{\sqrt{\tau^2-4}} + \frac{1}{2\tau}\right),$$

$$G' = \frac{\tau^2\left(\tau^2-2\right)\sqrt{\tau^2-4}}{2}, K = 3\pi\epsilon_0\epsilon_f\left(E_0 a\right)^2.$$

Now, we study the convergence of Eq. (4.2). We calculate $\log\left(\sum_{n=2}^{100}F_n^{(L)}/F_1^{(L)}\right)$ in four cases, which are shown in Figure 2. The results reveal that Eq. (4.2) is inapplicable when the two spheres are very close and κ is large. A further analysis indicates that when either of the following occurs, Eq. (4.2) is divergent and thus inapplicable: (i) $\kappa > 50$ and $\tau < 2.1$, (ii) $\kappa > 10$ and $\tau < 2.03$, or (iii) $\kappa > 4$ and $\tau < 2.004$.

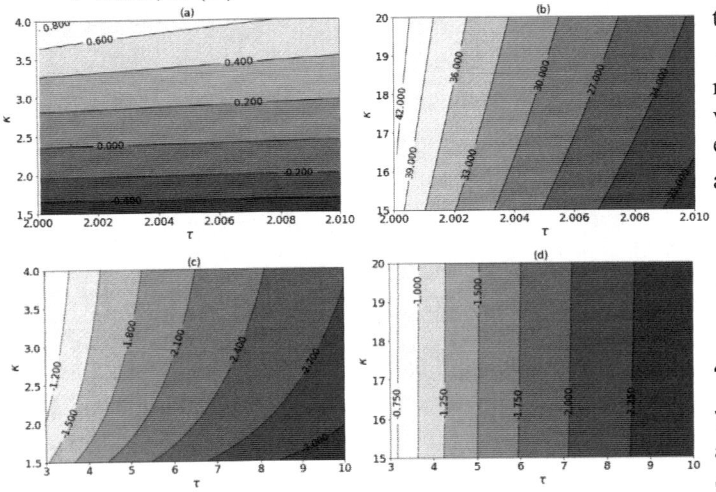

Figure 2. Contour line represents $\log\left(\sum_{n=2}^{100} F_n^{(L)} / F_1^{(L)}\right)$, as a function of the dielectric mismatch factor $\kappa = \epsilon_p / \epsilon_f$ and the spacing factor $\tau = d / a$. The solid lines represent positive values, while the dotted lines represent negative values. The values become stable when more $F_n^{(L)}$ terms are considered, except for the case in (b), where the values are not stable and diverge continually, since the values on the contour lines are far more than 1 and increase with the increase of n (When n comes to 580, the values on the domain of (b), *i.e.*, $(\tau, \kappa) = (2, 2.01] \times [15, 20]$, all exceed 100), which means Eq. (4.2) overestimates the dipole moment when the two spheres are very close and κ is large.

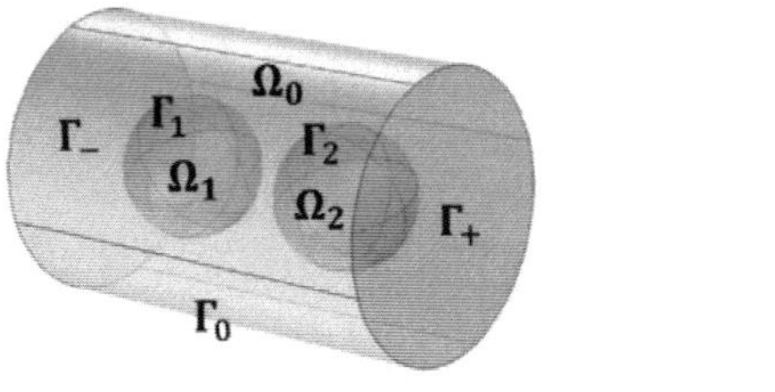

Figure 3. Schematic diagram of the model for a numerical calculation. Ω_1 and Ω_2 are two open bounded sets of two particles, and their boundaries $\Gamma_{1,2} = \partial\Omega_{1,2}$ are both Lipschitz continuous. $\Omega_0 = \overline{\Omega_1 \cup \Omega_2}$ is an open bounded set of the retaining cylinder. The Lipschitz-continuous boundary $\partial\Omega_0$ consists of the area of the top and bottom base $\Gamma_\pm$, the area of the side Γ_0, and $\Gamma_{1,2}$.

4.1 Model building

To ensure the accuracy of the numerical solution, axisymmetric structures are used in the model. As shown in Figure 3, to create an environment of a uniform electric field in the cylinder, different voltages $V_{+,-}$ are loaded on the top and bottom bases $\Gamma_{+,-}$. Two spheres are placed longitudinally with respect to the electric field in the cylinder. We define a as the radius of sphere, $\lambda_{\text{long}} a\tau + 2a$ as the height of the cylinder, and $\lambda_{\text{pol}} a$ as the radius of the cylinder. In this paper, two parameters λ_{long} and λ_{pol} are introduced to help determine a suitable mesh space that would be computationally inexpensive while ensuring that the assumption of a quasi-uniform electrical space is valid, which means we choose an arbitrary $(\lambda_m, \lambda_l, \lambda_p)$ with a small norm from the domain

$$\left\{(\lambda_m, \lambda_l, \lambda_p) \in \mathbb{R}_+^3 \middle| \max_{|\mathbf{v}|=1}\left\{\nabla_{\mathbf{v}} \tilde{W}_A(\lambda_m, \lambda_l, \lambda_p; \beta_0, \tau_0)\right\} < \varepsilon_0\right\},$$

with β_0, τ_0, and ε_0 fixed.

4.2 Method

The center of $\mathcal{A}$ is fixed along the longitudinal axis, and $\mathcal{B}$ is fixed at the origin. According to the principle of virtual work and Newton's third law, the scalar quantity of the force $\mathbf{F}$ that $\mathcal{A}$ exerts on $\mathcal{B}$ can be written as

$$F_{\mathcal{A}\to\mathcal{B}} = -F_{\mathcal{B}\to\mathcal{A}} = \frac{\delta W_{\mathcal{A}}}{\delta z} = \lim_{\Delta z \to 0} \frac{\int_{\mathcal{A}} w_{|d+\Delta z\rangle} d^3x - \int_{\mathcal{A}} w_{|d\rangle} d^3x}{\Delta z}, \quad (4.3)$$

where $w_{|d\rangle} = \frac{1}{2}\epsilon_0\epsilon_p \left|\mathbf{E}_{|d\rangle}\right|^2 = \frac{1}{2}\epsilon_0\epsilon_p \left|\nabla\phi_{|d\rangle}\right|^2$. $w_{|d\rangle}$, $\mathbf{E}_{|d\rangle}$, and $\phi_{|d\rangle}$ are, respectively, the electrostatic free energy density, the electric field and the potential function in the model where $\mathcal{A}$ is fixed at $z = d$. ϕ can be calculated using

$$\nabla^2\phi_i = 0, \text{ in } \Omega_i, i = 0,1,2, \quad (4.4)$$

with the boundary conditions

$$\begin{cases} \phi_0 = V_+ & \text{, on } \Gamma_+, \\ \phi_0 = V_- & \text{, on } \Gamma_-, \\ \phi_i = \phi_0 & \text{, on } \Gamma_i \quad , i = 1,2, \\ \kappa\dfrac{\partial\phi_i}{\partial\mathbf{n}_i} = \dfrac{\partial\phi_0}{\partial\mathbf{n}_i} & \text{, on } \Gamma_i \quad , i = 1,2, \end{cases} \quad (4.5)$$

where $\mathbf{n}_i$ denotes the normal to the boundary Γ_i, whose direction is from Ω_i to Ω_0, and the dielectric mismatch factor is $\kappa = \epsilon_p / \epsilon_f$.

Therefore, to determine the force, we simply calculate the distribution of the potential, where we apply a volume-discretization method, namely, the finite element method (FEM), due to its adaptability to complex geometries and its ease of handling discontinuous gradients of a variable. The basic steps of the FEM are shown in Figure 4. Based on the three-dimensional FEM in Sect. 5.3 of Ref. [28], we calculate the distribution of the potential, and therefore the electrostatic energy and the force.

To solve the problem beyond the regular form of particles and pay more attention to the post-processing, we use an unstructured grid for the FEM mesh. Tetrahedrons are employed as the cell shapes. The

Bowyer-Watson algorithm [29, 30] is used to compute the Delaunay triangulations. The parameters of the tetrahedral mesh generated using the Delaunay tessellation method are described below. The maximum element growth rate is 1.3. The curvature factor is 0.2. The average element quality is 0.7704. The subdivision parameter on the x,y,z-direction scale is λ_{mesh} for $\Omega_{1,2}$ and 1 for Ω_0. Figure 5 shows the FEM unstructured mesh used for the numerical calculation.

Solving the sparse matrix equation, the conjugate gradient method and the smoothed aggregation algebraic multigrid (SA-AMG) method are used [31], where the convergence criterion (the relative error after each iteration compared to that of the previous one) is set to be 1×10^{-3} to ensure the result converges well.

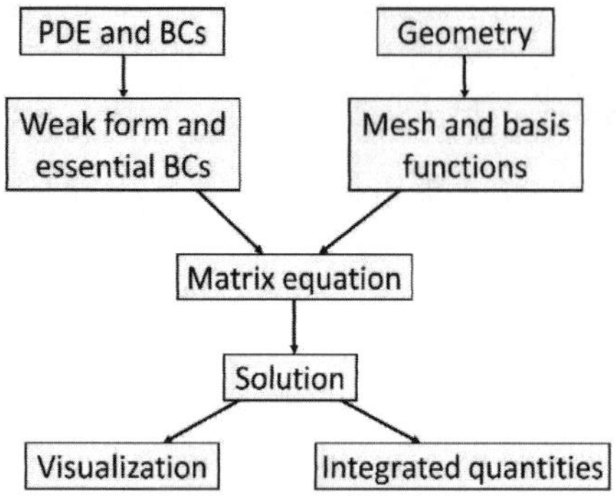

Figure 4. Basic steps to perform the finite element method (FEM) to solve a second-order partial differential equation (PDE) in a computational domain Ω with boundary conditions (BCs) on the boundary $\partial\Omega$ of the domain.

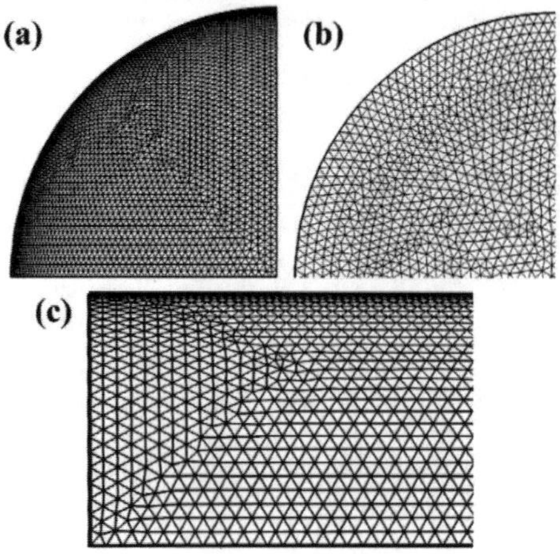

Figure 5. FEM unstructured mesh (a quarter of the whole part in view) used for the numerical calculation with $\lambda_{\text{mesh}} = 20$ (a) (y-z plan) for a spherical particle Ω_1, (b) (y-z plan) for the bottom of the cylinder, and (c) (x-y plan) for the side of the middle cylinder.

4.3 Electrostatic energy

The electrostatic energy of a single spherical particle in a uniform electric field can be calculated according to Eq. (1.4) as follows:

$$\mathbf{E}_{\text{in}} = -\nabla\phi_{\text{in}} = \frac{3}{\kappa+2}E_0\hat{\mathbf{z}} := E_{\text{in}}\hat{\mathbf{z}}, \tag{4.6}$$

$$\tilde{W}_A := \frac{W_A}{4\pi\epsilon_0\epsilon_f E_0^2 a^3} = \frac{\int_A w d^3x}{4\pi\epsilon_0\epsilon_f E_0^2 a^3} = \frac{3}{2}\frac{\kappa}{(\kappa+2)^2}. \tag{4.7}$$

Since $\tilde{W}_A{}' = -\frac{3}{2}\frac{\kappa-2}{(\kappa+2)^3}$, $\tilde{W}_A$ attains its maximum value at $\kappa = 2$.

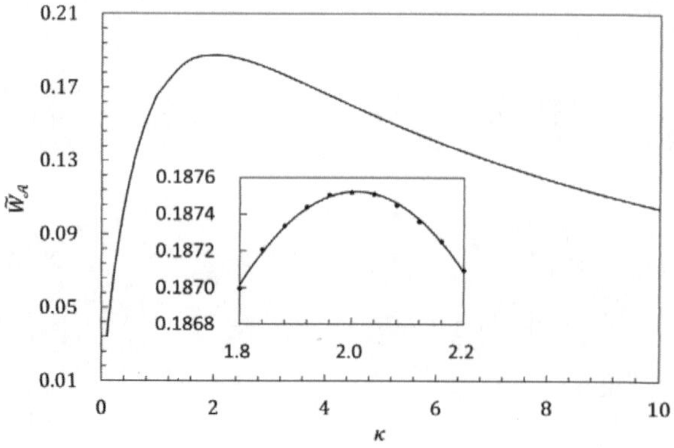

Figure 6. Electrostatic energy (in units of $1/(4\pi\epsilon_0\epsilon_f E_0^2 a^3)$) of the sphere in the single-sphere case where we consider a dielectric sphere of radius a and dielectric constant $\epsilon_0\epsilon_p$ being embedded in an electrically insulating liquid with dielectric constant $\epsilon_0\epsilon_f$ in response to an applied electric field $\mathbf{E} = E_0\hat{\mathbf{z}}$ as a function of the dielectric mismatch factor $\kappa = \epsilon_p/\epsilon_f$ ranging from 0.1 to 10.

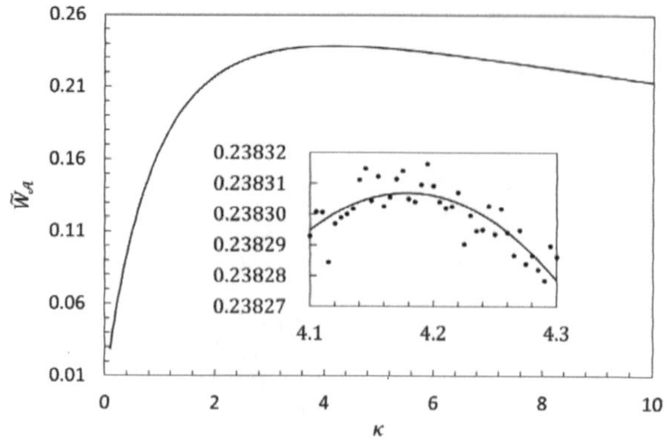

Figure 7. Electrostatic energy (in units of $1/(4\pi\epsilon_0\epsilon_f E_0^2 a^3)$) of each sphere in the double-sphere case where we consider two identical dielectric spheres, of radius a and dielectric constant $\epsilon_0\epsilon_p$ being embedded in an electrically insulating liquid with dielectric constant $\epsilon_0\epsilon_f$ in response to an applied electric field $\mathbf{E} = E_0\hat{\mathbf{z}}$, which is parallel to the line joining the centers of the spheres, as a function of the dielectric mismatch factor $\kappa = \epsilon_p/\epsilon_f$ ranging from 0.1 to 10.

In the simple-sphere case, τ is defined to be 2. As shown in Figure 6, for a series of test cases, the maximum relative error is 5.54×10^{-4} with $\kappa = 9.5$, which suggests the validity of the simulation.

As for the case of two spherical particles, the calculation becomes complex. In Figure 7, the dimensionless energy per sphere $\tilde{W}_A$ in the double-sphere case attains its maximum when κ is between 4.10 and 4.25. A similar trend occurs in the triple-sphere case. $\tilde{W}_A$ attains its maximum value of 0.8726 when κ is between 7.30 and 7.70.

4.4 Electrostatic force for the double-sphere case

Figure 8 presents the interaction energy of one particle $\tilde{W}_A$ for the double-sphere case as a function of the spacing d at κ=1, 2, 3, 5, and 10. Using Eq. (3.3), we obtain the curves shown in Figure 9.

We compare the numerical simulation result with the MID one. See Figures 9a-9d. When τ tends to 2, the condition $\kappa = 3$ is a criterion where the two results are approximately the same. When $\kappa < 3$, Eq. (3.2) underestimates the electrostatic force, while when $\kappa > 3$, Eq. (3.2) overestimates the force. When $\kappa \gg 1$ and τ tends to 2, there is a huge difference between the numerical results and the approximated theoretical results. For example, for $\kappa = 10$ and $\tau = 2.05$ the MID result is 617% that of the simulation. When τ tends to 2, all the MID results tend to infinity, which is impossible in physics, while the results from all simulation cases remain bounded.

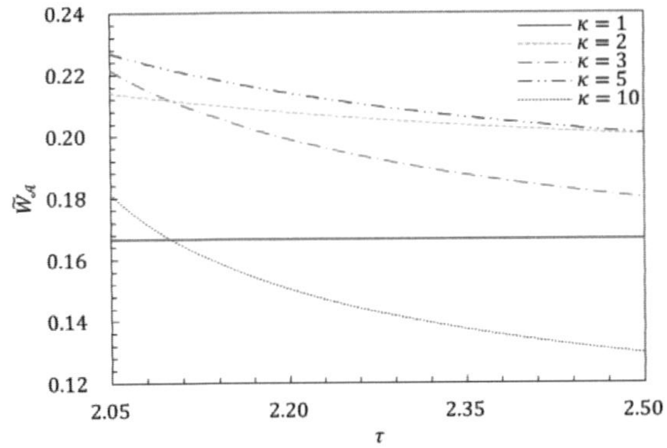

Figure 8. Electrostatic energy (in units of $1/\left(4\pi\epsilon_0\epsilon_f E_0^2 a^3\right)$) of each sphere in the double-sphere case, where we consider two identical dielectric spheres of radius a and dielectric constant $\epsilon_0\epsilon_p$ being embedded in an electrically insulating liquid with dielectric constant $\epsilon_0\epsilon_f$ in response to an applied electric field $\mathbf{E} = E_0\hat{\mathbf{z}}$, which is parallel to the line joining the centers of the spheres, as a function of the spacing factor $\tau = d / a$ ranging from 2.05 to 2.5 at κ =1, 2, 3, 5, and 10.

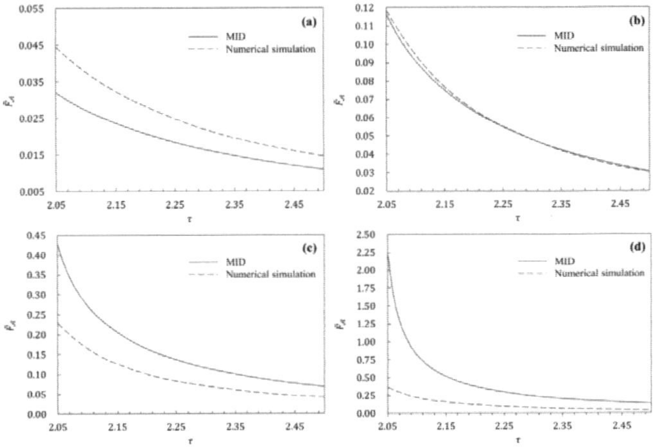

Figure 9. Electrostatic force (in units of $1/\left(4\epsilon_0\epsilon_f E_0^2 a^2\right)$) calculated in the MID (red solid line) *vs.* the numerical simulation (blue long dash line) of one sphere exerted by the other sphere in the double-sphere case as a function of the spacing factor $\tau = d / a$ ranging from 2.05 to 2.50 at (a) $\kappa = 2$, (b) $\kappa = 3$, (c) $\kappa = 5$, and (d) $\kappa = 10$.

5 Solution beyond the spherical form (three demonstrations)

5.1 Example 1: Earth-and-moon structure

In the first example, we consider two spheres with different radii. The sum of the volumes of the spheres is fixed and equals the volume of two spheres with the same radii a. Note λ is the ratio of the two radii. The radii can then be written as $\left(\frac{2a^3}{1+\lambda^3}\right)^{1/3}$ and $\lambda\left(\frac{2a^3}{1+\lambda^3}\right)^{1/3}$. In Figure 10, we plot the force $\tilde{F}_A = \frac{F_A}{4\epsilon_0\epsilon_f\left(E_0 a\right)^2}$ with different λ at $\kappa = 10$ and $d = (1+\lambda)\left(\frac{2a^3}{1+\lambda^3}\right)^{1/3}$. The maximum value of $\tilde{F}_A$ is reached when $\lambda = 1$. For $\lambda > 1$, $\tilde{F}_A$ decreases sharply, which reveals that to obtain the biggest force, the particles should be as identical as possible. The difference in size will result in great changes in the electrostatic force.

Now, we consider the sub-structures that are uniformly distributed across a sphere. To create the model, we consider N number of points distributed uniformly on a sphere, of which several approximate solutions have been proposed [32-34].

Consider $N = 20$. Since the regular dodecahedron has 20 vertices, its circumscribed sphere, the one that touches the regular dodecahedron at all vertices, can be considered as the base sphere. As shown in Figure 11, the Cartesian coordinates of the 20 vertices of a regular dodecahedron where the radius of its circumscribed sphere is $\sqrt{3}$ can be defined as follows:

$\left(0,\pm\frac{1}{\phi},\pm\phi\right)$, $\left(\pm\frac{1}{\phi},\pm\phi,0\right)$, $\left(\pm\phi,0,\pm\frac{1}{\phi}\right)$, $(\pm1,\pm1,\pm1)$,

where $\phi=\frac{1+\sqrt{5}}{2}=1.6180\cdots$ is the golden ratio.

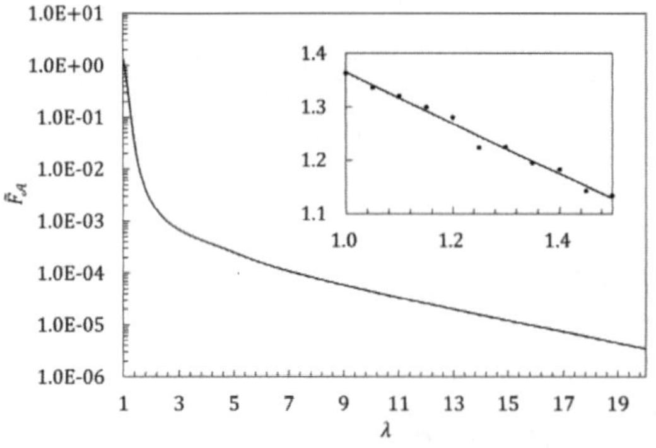

Figure 10. Semi-logarithmic plot of the electrostatic force (in units of $1/\left(4\epsilon_0\epsilon_f E_0^2 a^2\right)$) of one earth-and-moon structure as a function of the ratio of the two radii λ ranging from 1 to 20 at the different dielectric mismatch factor $\kappa=10$ and spacing factor $d=(1+\lambda)\left(\frac{2a^3}{1+\lambda^3}\right)^{1/3}$.

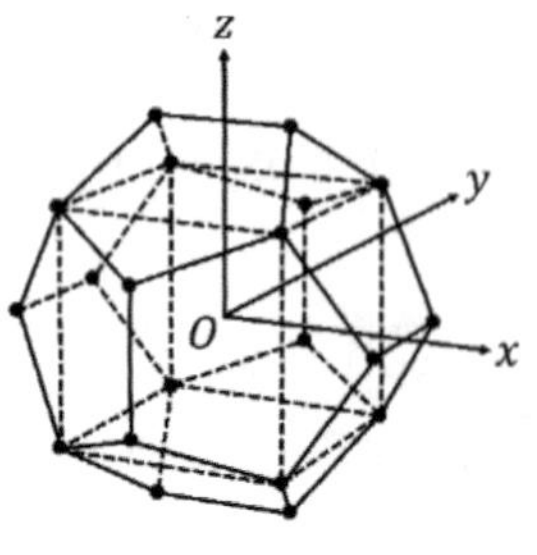

Figure 11. Regular dodecahedron in Cartesian coordinates

To rotate one of the particles, we introduce the rotation matrix. Given a unit vector $\hat{\mathbf{u}}=(x,y,z)$, the matrix for a rotation by an angle θ about an axis in the direction of $\hat{\mathbf{u}}$ is

$$\mathcal{M}(\hat{\mathbf{u}},\theta)=\exp\left(\begin{bmatrix}0 & -z\theta & y\theta\\ z\theta & 0 & -x\theta\\ -y\theta & x\theta & 0\end{bmatrix}\right). \quad (5.1)$$

In the following examples, we let $\hat{\mathbf{u}}=(x,y,z)=(0,0,1)$.

5.2 Example 2: Litchi structure

Consider two dielectric basic spheres $\mathcal{A}$ and $\mathcal{B}$ of identical radii a, with a distance d between them, on which there are 20 small spheres of radii $\lambda_{cp}a$ whose centers are situated on each of the 20 vertices of a regular dodecahedron as defined above for each basic sphere. We perform the rotation transformation $\mathcal{M}(\hat{\mathbf{u}},\theta)$ for one of the basic sphere; see Figure 12. The volume of one litchi-structured particle V_{cp} is calculated to be $\left(15\lambda_{cp}^4+40\lambda_{cp}^3+4\right)\pi a^3/3$. Consider $\lambda_{cp}=0.2$. Thus, $V_{cp}=1.448\pi a^3$.

Consider $\kappa=10$ and $\tau=d/a=2.1$. When $\theta=55°$, the two particles are nearly in contact with each other. Thanks to the symmetry, only the range $55°-90°$ needs to be studied. Figure 13 presents the electrostatic energy $\tilde{W}_A=\frac{W_A}{4\pi\epsilon_0\epsilon_f E_0^2 a'^3}$ as a function of θ ranging from 55° to 90°, where the effective radius a' is the radius of a sphere with volume V_{cp}. The energy declines with the increase of the rotation angle. Starting from 74°, the ratio between the energy at θ and that at 90° is smaller than 1×10^{-3}. Figure 14 shows the case for the rotation angle $\theta=\pi/2$. We calculate $\tilde{F}_A$ for the two-sphere case and the litchi structure using $\frac{F_A}{4\epsilon_0\epsilon_f\left(E_0 a\right)^2}$ and $\frac{F_A}{4\epsilon_0\epsilon_f\left(E_0 a'\right)^2}$, respectively. At the position in which the two particles are just in contact with each other, τ is 2 and approximately 2.08 for the two-sphere and litchi structures, respectively. It is noted that in both cases, when the two particles are increasingly near to each other, there is a significant increase in the force $\tilde{F}_A$, which tends to a finite value. The force at the limit position of the litchi structure is 129% that of the two-sphere case, which indicates that the litchi structure can help increase the electrostatic force.

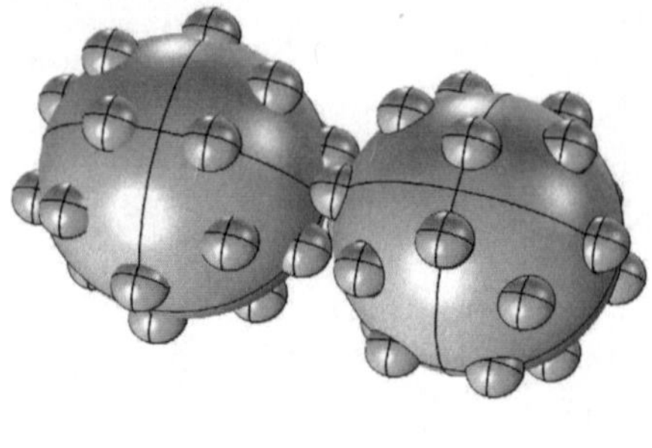

Figure 12. Schematic diagram of the litchi structure.

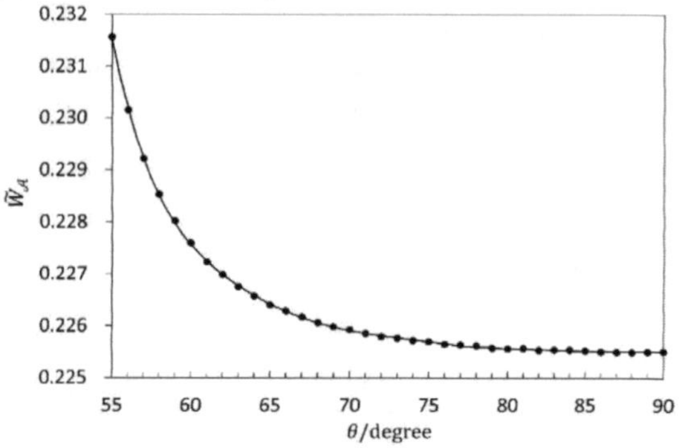

Figure 13. Electrostatic energy (in units of $1/\left(4\pi\epsilon_0\epsilon_f E_0^2 a'^3\right)$) of one litchi structure as a function of the rotation angle θ with

the dielectric mismatch factor $\kappa = 10$ and the spacing factor $\tau = d / a = 2.1$.

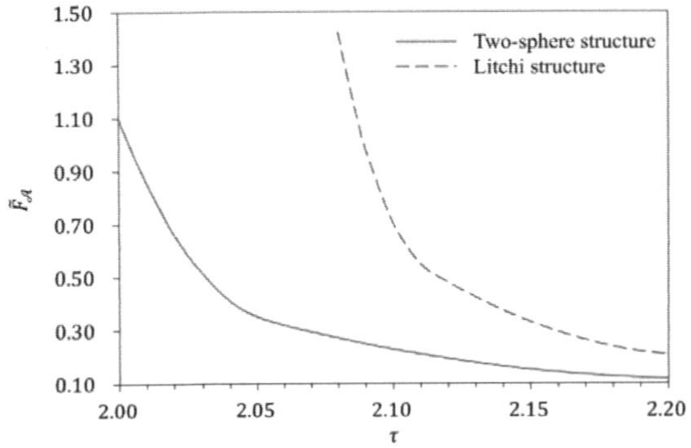

Figure 14. Electrostatic force (in units of $1/\left(4\epsilon_0\epsilon_f E_0^2 a^2\right)$ for the two-sphere structure and $1/\left(4\epsilon_0\epsilon_f E_0^2 a'^2\right)$ for the litchi structure) as a function of the spacing factor $\tau = d / a$ with $\kappa = 10$ for the two-sphere structure (red solid line) *vs.* the litchi structure (blue long dash line).

5.3 Example 3: Sea urchin structure

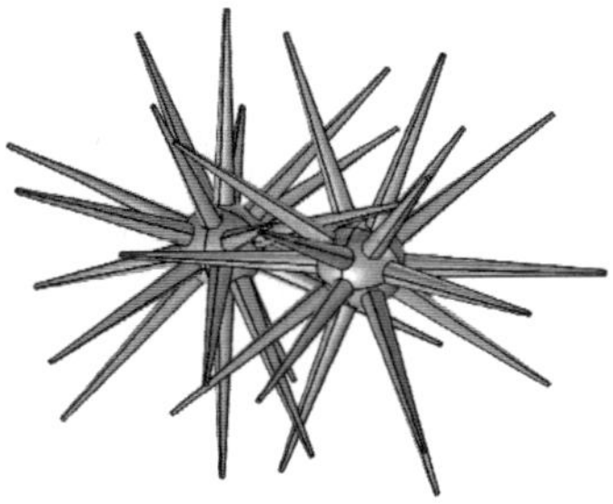

Figure 15. Schematic diagram of the sea urchin structure.

In this example, by replacing small spheres with spikes, we create a sea urchin structure, which is illustrated in Figure 15. For each sphere of radius a, 20 circular cone frustums with a height $\lambda_{\text{f,height}} a$ are uniformly distributed on it. The radius of the two bases equal $\lambda_{\text{bottom}} a$ and $\lambda_{\text{top}} a$, respectively. The distance between the sphere center and the top surface of a cone is $\lambda_{\text{p,height}} a$. Let $\lambda_{\text{f,height}} = 55/12$, $\lambda_{\text{bottom}} = 33/125$, $\lambda_{\text{top}} = 11/200$, and $\lambda_{\text{p,height}} = 11/2$. Therefore, parts of the frustums are inserted into the sphere. The volume of the above sea-urchin-structured particle equals a sphere with a radius $a' = 1.1702a$. Consider the rotation angle $\theta = 69°$ and $\kappa = 3$.

We calculate the electrostatic forces $\tilde{F}_A$ for the two-sphere case and the litchi structure using $\frac{F_A}{4\epsilon_0\epsilon_f (E_0 a)^2}$ and $\frac{F_A}{4\epsilon_0\epsilon_f (E_0 a')^2}$, respectively. As seen in Figure 16, unlike the litchi structure, the sea urchin structure contributes little to the electrostatic force and reduces the force: the force at the limit position of the sea urchin structure is 69% that of the two-sphere case. This is mainly because although this structure seems to be more closely integrated, in order to meet the assumption that no deformation occurs when the two particles are in contact with each other, the relative distance $\tau = d / a$ is quite limited (here τ should be larger than 3.35), which makes the force smaller than that in the two-sphere case where $\tau = 2$.

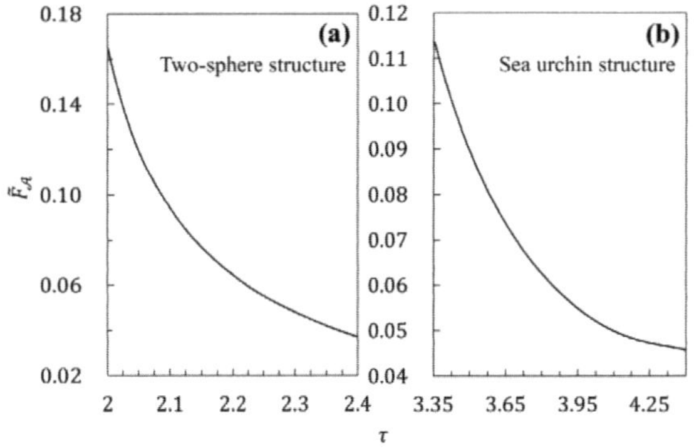

Figure 16. Electrostatic force (in units of $1/\left(4\epsilon_0\epsilon_f E_0^2 a^2\right)$ for (a) the two-sphere structure and $1/\left(4\epsilon_0\epsilon_f E_0^2 a'^2\right)$ for (b) the sea urchin structure) as a function of spacing factor $\tau = d / a$ with $\kappa = 3$ for (a) the two-sphere structure *vs.* (b) the sea urchin structure.

6 Conclusions

In this paper, we derive formulae of the electrostatic force between two spherical dielectric particles in the PD, DID, and MID models whose results are Eqs. (2.10), (3.7), and (3.8), respectively. The study of the convergence suggests that in the case of two identical spheres in MID models, the simplified version of Eq. (3.8), *i.e.*, Eq. (4.2), is inapplicable when either of the following occurs (i) $\kappa > 50$ and $\tau < 2.1$, (ii) $\kappa > 10$ and $\tau < 2.03$, or (iii) $\kappa > 4$ and $\tau < 2.004$. The numerical simulation based on the finite element method (FEM) is then necessary, by which we conclude the following:

- In the n-sphere case, the electrostatic energy per sphere at the limit position is a function of κ. When n is fixed, with the increase of κ, there exists a maximum energy.
- Compared to the case of having different particle sizes, the force between identical particles has the greatest force, which suggests making the particles as identical as possible in the experiment to increase the shear yield stress.
- Sub-structures uniformly grown on spherical particles can increase the force, but the effect depends on the balance of τ and the form of the sub-structures. A litchi structure is preferred.

Compared to the analytical methods, the FEM is a more efficient method to calculate the electrostatic force between particles, regardless of its form and relative permittivity. Thanks to this method, we could conceive a structure that can generate a greater force and create it in laboratory to obtain a better shear yield stress, which is applicable. Observed by an *in situ* electron microscope, we could build a 3D model of the particles and use the same method to calculate the force more accurately.

Acknowledgements

The project is fully supported by the personal savings of the corresponding author L. Gan who also undertook the bulk of the research. A great thanks to Y. H. Kwah, Y. T. Feng, G. Y. Mu, T. M. Nie, X. X. Han, and I. Zakia.

References

1. V. M. Zamudio, R. Nava, L. Rejón, M. A. Ponce, S. Víquez, and V. M. Castaño, Physica A: Statistical Mechanics and its Applications **227**, 55 (1996).
2. S. Choi, C. C. Cheong, and G. Kim, Mechatronics **7**, 53 (1997).
3. T. Kuriyagawa, M. Saeki, and K. Syoji, Precision Engineering **26**, 370 (2002).
4. N. Stevens, J. Sproston, and R. Stanway, Journal of mechanisms, transmissions, and automation in design **110**, 182 (1988).
5. Z. P. Shulman, R. G. Gorodkin, E. V. Korobko, and V. K. Gleb, Journal of Non-Newtonian Fluid Mechanics **8**, 29 (1981).
6. A. Simmonds, in *IEE Proceedings D-Control Theory and Applications* (IET, 1991), pp. 400.
7. N. K. Petek, (SAE Technical Paper, 1992).
8. R. Stanway, J. Sproston, and A. El-Wahed, Smart Materials and Structures **5**, 464 (1996).
9. K. Bohon, and S. Krause, Journal of Polymer Science-B-Polymer Physics Edition **36**, 1091 (1998).
10. H. See, Journal of Industrial and Engineering Chemistry **10**, 1132 (2004).
11. A. Lind, (FFI-rapport, 2008).
12. L. Marshall, C. F. Zukoski, and J. W. Goodwin, Journal of the Chemical Society, Faraday Transactions 1: Physical Chemistry in Condensed Phases **85**, 2785 (1989).
13. M. Parthasarathy, and D. J. Klingenberg, Materials Science and Engineering: R: Reports **17**, 57 (1996).
14. J. S. Marshall, and S. Li, *Adhesive particle flow* (Cambridge University Press, 2014).
15. K. Barros, and E. Luijten, Physical Review Letters **113**, 017801 (2014).
16. K. Barros, D. Sinkovits, and E. Luijten, Journal of Chemical Physics **140**, 064903 (2014).
17. J. M. Kosterlitz, and D. J. Thouless, Journal of Physics C: Solid State Physics **6**, 1181 (1973).
18. R. Tao, J. Woestman, and N. Jaggi, Applied Physics Letters **55**, 1844 (1989).
19. W. Wen, X. Huang, and P. Sheng, Soft Matter **4**, 200 (2008).
20. S. Gawade, and A. Jadhav, International Journal of Engineering Research & Technology **1**, 1 (2012).
21. A. Goyette, and A. Navon, Physical Review B **13**, 4320 (1976).
22. R. Tao, Q. Jiang, and H. Sim, Physical Review E **52**, 2727 (1995).
23. Y. Siu, J. T. Wan, and K. Yu, Computer Physics Communications **142**, 446 (2001).
24. J. D. Jackson, *Classical Electrodynamics*, 3rd ed. (Wiley, 1999).
25. J. N. Foulc, P. Atten, and N. Felici, Journal of electrostatics **33**, 103 (1994).
26. K. Ahn, and D. J. Klingenberg, Journal of Rheology **38**, 713 (1994).
27. J. Cao, J. Huang, and L. Zhou, Journal of Physical Chemistry B **110**, 11635 (2006).
28. J. M. Jin, *The Finite Element Method in Electromagnetics*, 3rd ed. (Wiley, 2015).
29. D. F. Watson, The Computer Journal **24**, 167 (1981).
30. A. Bowyer, The Computer Journal **24**, 162 (1981).
31. P. Vaněk, J. Mandel, and M. Brezina, Computing **56**, 179 (1996).
32. M. Sibuya, Annals of the Institute of Statistical Mathematics **14**, 81 (1962).
33. R. L. Smith, Operations Research **32**, 1296 (1984).
34. E. B. Saff, and A. B. Kuijlaars, The Mathematical Intelligencer **19**, 5 (1997).

Scale-Up of Solid-Liquid Mixing Based on Constant Power/Volume and Equal Blend Time Using VisiMix Simulation

Megawati[1], Bayu Triwibowo[1], Karwono[2], Waliyuddin Sammadikun[1] and Rofiatun Musfiroh[1]

[1]*Chemical Engineering Department, Engineering Faculty, Universitas Negeri Semarang, Semarang, Indonesia*
[2]*Mechanical Engineering Department Engineering Faculty, Universitas Negeri Semarang, Semarang, Indonesia*

Abstract. Mixing is one of the important process in many areas of chemical industries, for instance pharmaceutical, drug, ink, paint and other industries. Solid-liquid suspension is produced for 80% of all mixing industries such as leaching process, crystallization process, catalytic reactions, precipitation, coagulation, dissolution and other applications. Two main objectives in solid-liquid mixing namely, avoid settling of solid particles on the tank bottom and ensure the solid particles are uniformly distributed. Many factors that can affect the quality of solid-liquid mixing, they are tank geometry, impeller geometry and speed, baffles, viscosity and density of media. Scale-up of the process is important to conduct before produce it on commercial scale. Two parameters for scale-up solid-liquid mixing are equal blend time and power per volume. Before scaling up the process to industrial scale, an engineer must know the condition of the mixture between both of two. VisiMix can simulating scale-up of solid-liquid mixing in order to know the phenomena inside the tank without conducting a large number of experiments and cheaper. The simulation start from keep the ratio of impeller to tank diameter remains constant, then change the condition operation of mixing. In this paper, power per volume parameter is more recommended as a result of the degree of uniformity of solid phase in liquid.

1. Introduction

Mixing in stirred tank is one of important process in many areas of chemical industries, for instance pharmaceutical, drug, food, ink, paint and other industries. Mixing is a dynamic generated by impeller rotation where the difference of pressure in various areas of the stirred tank acts as driving force [1]. The object of mixing is to reduce the heterogeneous properties within the tank and provide a uniform mixture. Also, mixing give a better mass transfer in the shortest possible time [2].

Stirred tank was equipped by one or multi impellers like propeller, paddle, or turbine. The impeller creates a flow circulation and mix the fluid within a stirred tank to cause bulk motion in micro scale. The motion is promoted by transfer energy of the impeller into the process of fluid [3]. The process fluid may be single phase (miscible or immiscible liquid) or multiphase (solid, liquid and gas).

Solid-liquid suspension is produced for 80% of all mixing industries such as leaching process, crystallization process, catalytic reactions, precipitation, coagulation, dissolution and other applications [4]. There are two main objectives in solid-liquid mixing namely, avoid settling of solid particles on the tank bottom and ensure the solid particles are uniformly distributed. In many operations, it is essential to keep all solids in motion in order to prevent the accumulation of solids on the tank. Another important objective of solid-liquid performance is the distribution of solids throughout the tank. Very often, a uniform dispersion of solids throughout the tank is necessary to ensure adequate exposure to the process conditions. However, quantitative information in this area is limited and mostly are confined to low concentration in small tank [3].

To get two of those objectives, the speed of impeller rotation should be higher than Just Suspension Speed (JSS). JSS is the minimum impeller speed that does not result in settling of solid particles. The higher impeller speed will increase the uniformity of solid phase. Therefore, the suspensions of solid particles in liquid media must be complete in order to meet a requirement set in the industry.

Many factors can affect the quality of solid-liquid mixing, they are tank geometry, impeller geometry and speed, baffles, density and rheological properties of liquid. Those factors are a requirement in designing stirred tank, for example the position of three blades propeller within the liquid is 1/3 of the media level [5].

Scale-up is important to conduct in order to produce a chemical product on commercial scale. A good quality of mixing-based product depends on condition of mixing itself on commercial-scale. The product is tried in lab-scale then study the performance of mixing process on it. Two parameters for scale-up are mixing time and power.

These parameters must be close as possible both of these scales by arranging impeller, tank geometry, and operating condition. The condition within a stirred tank can be known by VisiMix Simulation software. This paper will discuss scale-up solid-liquid mixing including axial and radial distribution of solid phase, energy and force, macromixing time and hydrodynamic output. Finally, an engineer can be aware some consideration in order to make mixing-based product on industrial scale.

2. Theory

2.1. Complete suspension

A suspension is defined as being completed when no particles remain on the tank bottom for more than 1-2 s [6]. Under this condition the total surface of the solid phase are suspended in the liquid and will cause mass transfer phenomena.

A solid particle is just lifted from the bottom of a stirred tank when the forces exerted on that particle by the turbulent motion of the fluid become equal to apparent weight of that particle where the solid particles size is between 100 μm and 2 mm [7]. For normal mixing operation the latter length scale is 40 μm orders and determined by the impeller dimensions. So, it may assume that the size of the eddies which pick up the particles is in the inertial subrange of the turbulence spectrum [8].

2.2. Just suspension speed

Just Suspension Speed (JSS) is the minimum impeller speed at which no solid particle remains stationary on the tank bottom for ≤ 1-2 second and with the accuracy of 2-3% [9]. Just suspension speed could be estimated by Zwietering correlation. The correlation based on experiments and its dimensionless number as is expressed in Equation 1.

$$N_{js} = sv^{0.1}\left(\frac{g\Delta p}{pl}\right)^{0.45} d_p^{0.2} X^{0.13} D^{-0.85} \quad (1)$$

Where N_{js} is just suspension speed, s is the geometrical constant, υ is kinematic viscosity, g is gravity acceleration, Δp is density difference, *pl* is liquid density, d_p is particle size, X is mass ratio of solid to liquid, and D is impeller diameter.

2.3. Scale-up of solid-liquid mixing

Equal blend time and power per volume (P/V) are parameters that frequently used to increase the capacity of mixing-based product. Impeller can operate to mix a mixture system within a stirred tank due to a power. Equal blend time consider the time required to achieve the maximum of the degree uniform mixture. An engineer must clear what the need of mixing process. For instance, equal blend time must be selected if the process is faster [10].

In case where the impeller speed higher than JSS, scale-up should be done on P/V criteria. In the other research [11] stated that when similar P/V is maintained of two units that geometrically similar, the solid-liquid mass transfer coefficient remains similar.

2.4. Geometry similarity

Geometry similarity tank for P/V rules can be used when the density difference between solid and liquid is small [12]. The design of mixing tank both lab-scale and commercial scale are shown in Figure 1. Equation 2 is satisfied for exact geometry similarity [10].

$$C_1/T_1=C_2/T_2;\ D_1/T_1=D_2T_2;\ W_1/T_1=W_2/T_2;\ Z_1/T_1=Z_2/T_2 \quad (2)$$

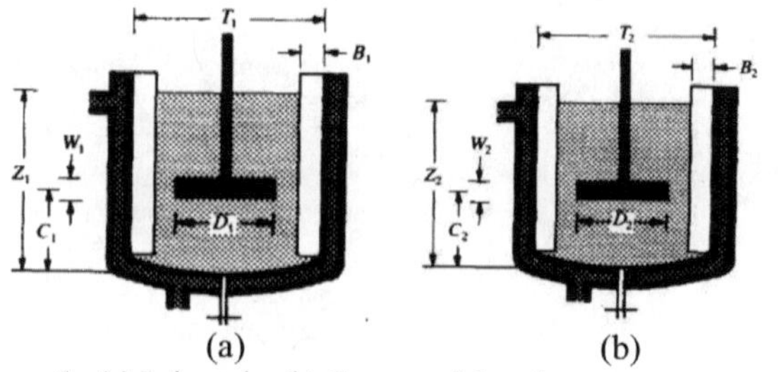

Figure 1. (a) Lab-scale (b) Commercial-scale.

2.5. VisiMix simulation

VisiMix is a simulation application for many mixing processes. The model of simulation based on Kolgomogoroff, Hinze, and Levich applied hydrodynamics approach. VisiMix bridges the gap that separates engineering practice from achievement in the field of mathematical simulation. VisiMix provides chemical and process engineer to perform technical calculations based on the results of the most recent experimental research and mathematical modeling of mixing process [13].

3. Methods

3.1. Data source

The data used in this paper was obtained from Kuzmanic and Ljubicic [14]. There were 320 mm of tank diameter, impeller diameter of 106 mm (the speed of 500 RPM), ratio liquid height to tank diameter of 1, 32.2 mm of four baffles width, The measurement were conducted using polyethylene particles (ρ = 840 kg/m^3) and particles diameter of 200-600 μm, 25.72 L of water at 15°C (density = 997 kg/m^3, viscosity = 1,4 x 10^{-3} Pa s). Reynolds number at this speed can be calculated by Equation 3,

$$Re = \frac{ND^2\rho}{\mu} \quad (3)$$

where N is impeller speed, D is impeller diameter, ρ is density, and μ is viscosity. So the Reynolds number was 66878. The parameter of N_p can be calculated by Equation 4,

$$P = N_p n^3 D^5 \rho \quad (4)$$

where P is mixing power, ρ is density, n is impeller rotation.

3.2. Simulation of mixing process

VisiMix Turbulent SV software was used to simulating mixing tank on lab-scale and industrial scale. The parameters are power per volume (P/V) and mixing time. P/V rule for geometrically similar tank can be used when the density difference between solid and liquid is small [10]. Besides that, the similarity in solid-liquid mass transfer coefficient can be achieved as power per unit volume parameter is maintained. Design of industrial-scale tank will be adjusted by changing speed and diameter of impeller and diameter of tank. Propeller impeller used in this simulation, to compare hydrodynamic characteristic. The ratio of D/T was 1/3. The data of physical property of liquid obtained to the research [14].

4. Results and discussions

In this simulation, we use the results of Kuzmanic Ljubicic [14] as key process parameters with the results shown in Figure 2 and Figure 3.

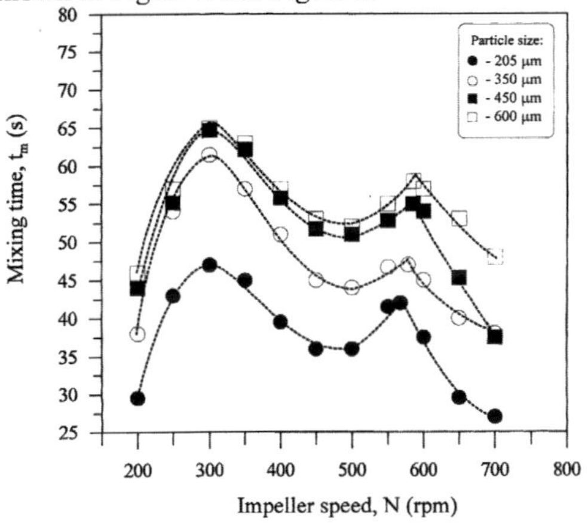

Figure 2. Effect of particle size on the mixing time [14].

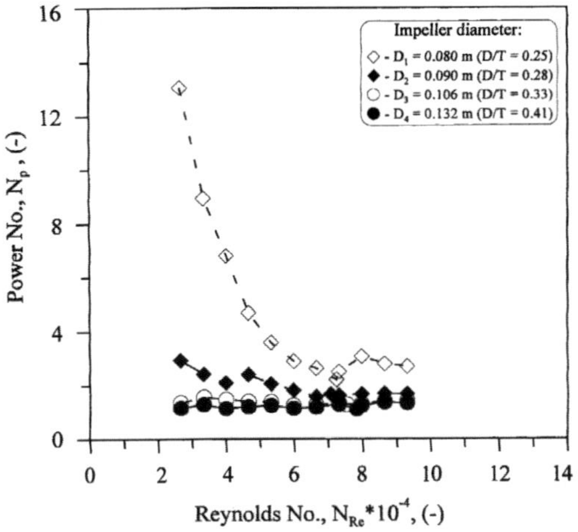

Figure 3. Effect of Reynolds number power number for the different impeller diameter [14].

According to Figure 2 the mixing time value was 36 s. Figure 3 shows the power number was 1.5 so the mixing power was 11.44 W.

4.1. Scale-up geometry similarity

The geometry of industrial-scale in this study was 6000 L. According to Figure 1, the tank dimensions are listed in Table 1.

Table 1. Geometry similarity between lab scale and industrial scale.

Geometry	Lab Scale	Industrial Scale
Tank diameter	0.32 m	2 m
Impeller diameter	0.106 m	0.668 m
Impeller clearance	0.106 m	0.668 m
Height of liquid	0.32 m	2 m
Baffle width	0.032 m	0.2 m

4.2. Simulation of P/V parameter

The power required in mixing process is the key factor to design and operate an impeller. Also, this factor can help an engineer to estimate mixing performance. Therefore, it must be considered that the power in lab-scale and industrial-scale remain similar according to each volume.

Unit volume for mixing tank must be known before determined the ratio of power per volume with Equation 5.

$$V = \frac{\Pi T^3}{4} \quad (5)$$

According to Equation 5, the volume of lab-scale tank was 0.0257 m^3 . So the P/V value was 445. The value will be a parameter in this simulation.

The impeller speed will be specify when P/V value are as close as possible both two scale. Thus, the speed of impeller rotation was 242 RPM. The simulation results can be seen in Table 2.

Table 2. Simulation result based on P/V.

Output Calculation	Units	Value
Mixing Power	W	2820
Torque	N*m	111
Force applied to impeller blade	N	111
Force applied to baffle	N	23.9
Impeller tip velocity	m/s	8.53
Macromixing time	s	27.5
Depth of Vortex	M	0.103

Using Equation 5, the unit volume in industrial scale is 6.33 m^3. The power in this simulation was 2820 watt. So the P/V value was 445.49. The value is close to the P/V in lab-scale. Therefore, the design of the tank and the impeller on industrial-scale are satisfied.

Sedimentation of solid particles on the tank bottom must be prevented in order to reaches a good quality of liquid-solid mixing. So the impeller speed in this simulation must be higher than Just Suspension Speed (JSS). According to the results of liquid-solid mixing calculation, the pick up conditions is satisfied and the impeller speed is higher than JSS.

The solid distribution of this simulation are shown in Figure 4 and Figure 5. The corresponding value of axial and radial non-uniformity are 4.54 and 0.189 respectively.

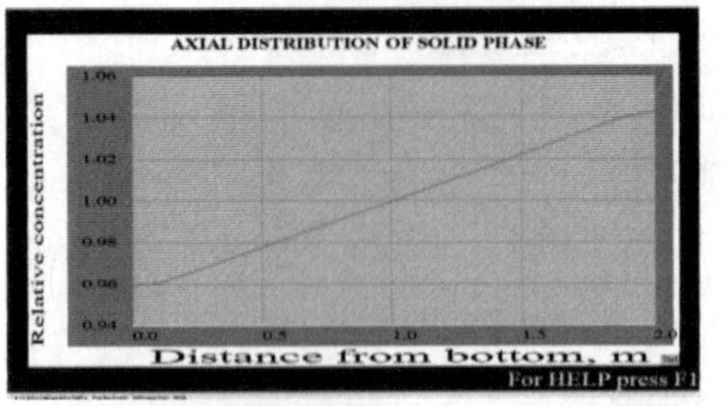

Figure 4. Axial distribution of solid phase in P/V parameter.

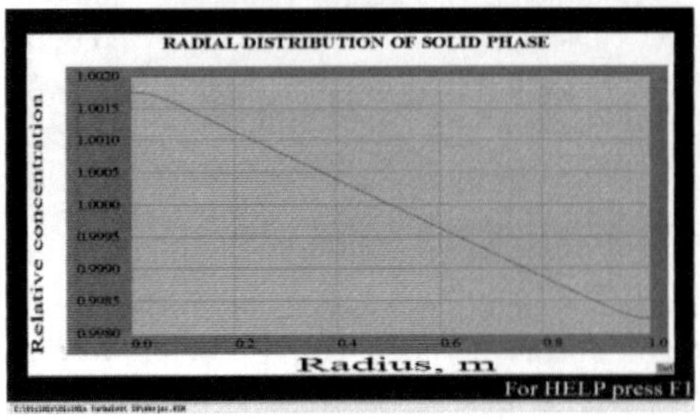

Figure 5. Radial distribution of solid phase in P/V parameter.

4.3. Scale up based on mixing time

Mixing time is the time required to get a maximum degree of uniformity. It can be predicted by knowing the tank size, impeller design and operating condition, and properties of liquid. Efficiency a mixing process can be known by mixing time. There are two type of mixing time scale, namely micromixing and macromixing [15]. This simulation shows that on 110 RPM of impeller speed, the mixing time was 35.5 s. Table 3, Figures 6 and 7 show the result of the simulation with the same of geometry.

Table 3. Results of simulation based on mixing time.

Parameters	Units	Value
Mixing Power	W	268
Torque	N*m	23.3
Force applied to impeller blade	N	23.2
Force applied to baffle	N	6.47
Impeller tip velocity	m/s	3.85
Macromixing time	S	35.5
Depth of Vortex	m	0.0127

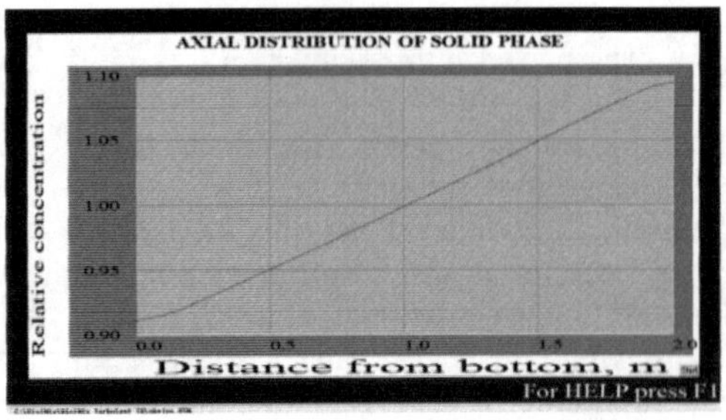

Figure 6. Axial distribution of solid phase in equal blend time parameter.

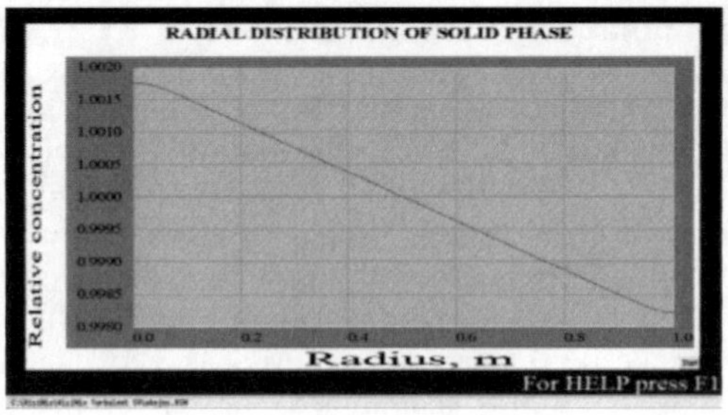

Figure 7. Radial distribution of solid phase in equal blend time.

The corresponding value of axial and radial non-uniformity are 10.1 and 0.189 respectively. Distribution of solid phase in industrial scale based on P/V is more uniform than mixing time based parameter.

4.4. Comparison between P/V and equal blend time

Table 4 shows the comparison of simulation results between P/V and equal blend time.

Table 4. Comparison between P/V and equal blend time parameter.

Simulation Output	Parameters	
	P/V	Equal Blend Time
Torque (N*m)	111	23.3
Mixing Power (W)	2840	268
Macromixing Time (s)	27.5	35.5
Axial nonuniformity (%)	4.54	10.1
Radial nonuniformity(%)	0.189	0.189

From Table 4, it can be seen that the axial distribution of solid phase in P/V parameter is more uniform and faster for complete suspension. But, P/V parameter gives the higher energy required. Usually equal blend time gives the higher energy than P/V parameter because the increasing of impeller speed. It happened because the impeller used in this simulation is different with the research [13]. So in this process, it is better to use propeller rather than pitched blade turbine because with the same of quality of mixing, propeller give a faster process.

5. Conclusion

In this paper, power per volume parameter is more recommended as a result of axial non-uniformity (4.54) lower than mixing time parameter (10.1). When the mixing power and cost becomes a consideration, scale-up using mixing time parameter must be chosen due to the power needed of 268 W, much lower than the P/V parameter (2840 W).

Acknowledgment

The authors are indebted to Ministry of Research, Technology and Higher Education of the Republic of Indonesia for financial support via RAPID grant 2018.

References

1. F. Strek, *Mixing and stirring apparatuses (Mixer)*, (WNT, Warsawa,1981)
2. C. Kuncewicz, K. Szulc, and T. Kurasinski, Chem. Eng. Process. **44**, 766-774 (2005)
3. A.T. Mak, Thesis, University of London (1992)
4. M. Spidla, M. Mostek, V. Sinevic V, M. Jahoda, V. Machon, Chem. Pap. **59**,386-393 (2005)
5. S.M Walas, *Chemical Process Equipment Selection and Design* (Department of Chemical and Petroleum Engineering, University of Kansas, 1990)
6. R.K. Geisler, C. Buurman, A.B. Mersmann, Chem. Eng. J. **51** 29-39 (1993)
7. C. Buurman, G. Resoort, A. Plaschkes, Chem. Eng. Sci. **41**,2865-2871 (1986)
8. J.O. Hinze, *Turbulence, 2nd edn*. (McGraw-Hill, New York, 1975)
9. Th. N. Zwietering, Chem. Eng. Sci. **8**, 244-253 (1958)
10. R.E. Brodkey and C. H. Harry, *Transport phenomena: A unified approach* (McGraw Hill International, New York 1998)
11. V. Koganti, F. Carrol, R. Ferraina, R. Falk, Y. Waghmare, M. Berry, Y. Liu, K. Norris, R. Leasure, and J. Gaudio, Pharm. Sci. Technol. **11(4)**, 1541-1548 (2010)
12. D.M. Levins, J.R. Glastonburry, Chem. Eng. Sci. **27**,537-543 (1972)
13. VisiMix, "Scientific Backgorund", (2018). Available: http://visimix.com/support/scientific-background/. [Accesed: 01-April-2018]
14. N. Kuzmanic, B. Ljubicic, Chem. Eng. J. **84,** 325-333 (2001)
15. A.R. Rao, B. Kumar, Chin. J. Chem. Eng. **17(2)**, 355-358, (2009)

AUTHOR INDEX